FIVE PHOTONS

FIVE PHOTONS

REMARKABLE JOURNEYS OF LIGHT ACROSS SPACE AND TIME

JAMES GEACH

REAKTION BOOKS

For Sophie and Lucy,
just starting their journey.

Published by
REAKTION BOOKS LTD
Unit 32, Waterside
44–48 Wharf Road
London N1 7UX, UK
www.reaktionbooks.co.uk

First published 2018

Printed and bound in Great Britain by
TJ International, Padstow, Cornwall

A catalogue record for this book is available from the British Library

ISBN 978 1 78023 991 0

CONTENTS

ONE

WHAT IS LIGHT?

When I got up this morning it was still dark outside. Groggy with sleep, I negotiated my way downstairs, instinctively treading the steps and running my hand down the wall. At the bottom of the stairs I felt for and flipped a switch. Almost instantly, light raced out from two bulbs on the ceiling. Within a few billionths of a second that light started smashing into all the various things in the room: the wood laminate floor, the yucca plant in the corner, the sofa, the television, my daughters' toys scattered here and there.

Some of the light hitting all that domestic paraphernalia bounced back towards me. Then, about ten nanoseconds after it left the bulb, a small fraction of that reflected light was intercepted by two small apertures in my head, passing through a couple of slightly squashy, transparent and rather usefully shaped blobs of biological tissue on the way. This subtly altered the direction of the light's path, focusing it on certain photosensitive cells at the back of my eyes and triggering a response that was sent down my optic nerve to my brain through a bioelectrical stimulus. Quickly decoding this stream of information, my visual cortex rendered the scene for me. I could see the room.

Our experience of light is simple and intimate: we can see it. Light interacts with objects in our environment in different ways, and this allows us to distinguish, say, an antique oak table from a plastic chair, and polished metal from fur. Colour is maybe the most obvious example of this interaction. The bulbs in my living room emit 'white light', which is a mixture of every colour of the rainbow. In physics we'd call it a broad spectrum light source. But the things I see around me that are illuminated by that white light have different colours and shades. So how do we go from a broad spectrum source like a light bulb to the huge range of colours we see in the environment illuminated by that source?

The answer is in the way different materials absorb and reflect the incident light. If an object appears blue then it is absorbing all the colours in the white light *except* for blue, which is reflected. If an object appears totally black, then it is reflecting none of the light hitting it. White surfaces reflect all of the incident light.

We are so familiar with the stuff, but what *is* light? How does it travel from one spot to another? What does it actually mean for light to have colour, or to be reflected or absorbed by an object? Like many simple questions about the world we live in, these run deep.

In physics, when we first start learning about light and its properties, we usually begin with the so-called 'laws' of optics. These are the rules of how light travels through and between different media, such as air, glass or water. We think about light 'rays' that travel in straight lines from their origin as they traverse space. The behaviour of those rays can be described using some fairly simple rules. I'll give you an example. Take a flat mirror. A ray of light hitting the mirror will be reflected off the silvered surface, bouncing back. This we know pretty well. But the reflected ray doesn't bounce off the mirror randomly, it bounces off at an angle exactly the same as the incoming ray relative to the 'normal', which is an imaginary plane perpendicular to the reflective surface. This is called the law of reflection.

There are other effects that we encounter every day. Take a light ray passing from one medium to another, for example. If you have ever put a straw in a glass of water, you'll have noticed that the straw appears to be bent, or disjointed, where it enters the water. Of course, the straw itself has not actually broken, but your eye perceives it as so because light rays from the submerged part have to travel through the water, then into the air towards your eyes. Compare this journey to that of the light rays coming from the top of the straw sticking out of the water: they only have to travel through the air to get to your eyes.

As a ray encounters a new medium, it can change direction slightly. This is called refraction. The reason the ray alters direction is because the speed of light can change in different materials. Yes, the speed of light is a fundamental constant, but that refers to light travelling in a pure vacuum. In glass, for example, light travels at about two-thirds of its maximum speed. You can see an analogous effect in water waves: a set of parallel waves approaching a shore can change direction depending on the depth of the water. Since ocean waves travel faster in deeper water, any part of the wave passing over shallow water – say, a sandbar – will slow down and lag behind, introducing a bend in the wave front. The direction of propagation of the waves is deflected.

In our straw-in-glass example, what you *perceive* as the source of the light – that is, where your brain interprets an object to be – is the apparent straight-line origin of each ray that enters your eye. Since the rays coming from below the water line get deflected as they pass from water to air, it appears that the submerged bit of the straw is slightly displaced compared to where you expect it to be.

Refraction also explains rainbows. If I hold up a chunk of glass to the sunlight, it might cast a rainbow on the wall. The

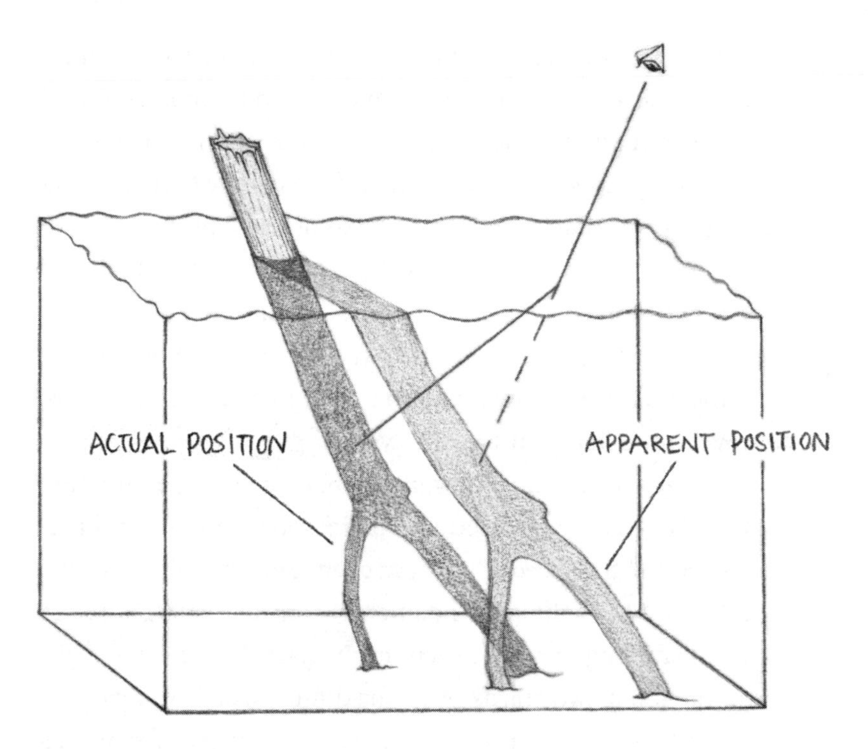

Refraction

When light passes from one medium to another, its speed can change. This alters the direction of propagation of the ray, called refraction.

refractive index of the glass, or the amount that light slows down as it passes through it, is slightly different for different colours of light. Sunlight, like our light bulb, is made up of a broad spectrum of colours, and so you can think of sunlight as a collection of different rays of light, each a different pure colour. When sunlight is refracted, the colours get dispersed into a rainbow because each ray is deflected by a slightly different angle, and the size of that angle depends on the colour.

Those were some phenomenological descriptions of the behaviour of light. If you wanted to design an optical system – say a pair of spectacles – then these rules would serve you pretty well. But they don't describe exactly *what* light is. Can we dig a little deeper? I used an ocean-wave analogy above for good reason: light is a type of wave. It's an *electromagnetic* wave.

What does this mean? Electromagnetism is one of the four fundamental forces of nature, sitting alongside the force of gravity and the 'strong' and 'weak' nuclear forces. Those nuclear forces govern the structure of the nuclei of atoms, the building blocks of the material world, and they operate over extremely short distances. Gravity, as we all know, is the attractive force between any two objects in the Universe with mass, and can act over infinite range. It is the force that holds us onto the Earth, keeps the Earth in orbit around the Sun, and generally determines the overall distribution of mass in the Universe, from solar systems to clusters of galaxies. Electromagnetism is a force that acts between particles with 'charge'.

Charge is a fundamental property of subatomic particles: a particle can either be positively charged, negatively charged or have zero charge (we call that 'neutral', but we will ignore those for now). Particles with the same charge repel each other, and particles with opposite charges are attracted. A good analogy is the attraction and repulsion of the poles of a magnet. Like the force of gravity, the strength of the electromagnetic force scales with the separation of the charged particles, following an 'inverse square law': double the distance between two charged particles and the strength of the force between them drops by a factor of four. Halving the distance increases the strength of the force by a factor of four, and so on.

An example of a positively charged particle is a proton. Protons and neutrons (subatomic particles with zero charge) make up the dense nucleus of an atom. Actually, protons and neutrons are each made from groups of three subatomic particles called quarks, but we won't go down that rabbit hole here. Suffice it to say, for now, we can think of protons and neutrons as distinct particles. You may wonder, if the electromagnetic force repels particles with the same charge, how does the nucleus stay together? Shouldn't all those protons repel each other to disastrous effect? That's where the strong nuclear force comes in: it acts over very short distances to glue protons and neutrons together, and is stronger than the proton–proton repulsion on those small scales. So, overall, the nucleus of an atom can be considered a positively charged particle, where the total positive charge is set by the number of protons in the nucleus.

Surrounding the positively charged nucleus are the electrons, typically in equal number to the protons. Electrons are another type of subatomic particle, but negatively charged, and about two-thousandths of the mass of a proton. A proton carries a charge of +1, and an electron carries a charge of –1. Their charges balance. So the net charge of an atom – the sum of the charges of the electrons and protons – is zero, or neutral. By successively removing electrons from an atom, which can be done by giving an electron enough energy to break free from the attraction of the nucleus, the atom can become positively charged overall. We call this process 'ionization' and we refer to the electron-stripped atom as an ion. Conversely, negatively charged ions are formed by introducing extra electrons to an atom.

The force between charged particles plays a crucial role in nature: it binds atoms together into larger structures called molecules. The fact that you can't push your thumb through the palm of your hand is because of the 'electrostatic' bonds between molecules in your skin, muscle and bone.

A simple example of the formation of such bonds can be found in common salt. Salt is just the colloquial term for the molecule sodium chloride, which fundamentally is comprised of pairs of sodium and chlorine atoms. These atoms bond in the following way: under certain conditions, sodium can donate one of its electrons to the chlorine atom, each becoming an ion in the process. The sodium has a net positive charge because it has lost an electron, and the chlorine has a net negative charge because it has gained one. The two oppositely charged ions attract each other, and we call this an ionic bond. Countless sodium and chlorine atoms can bond together this way, arranging themselves into a regular lattice that forms salt crystals.

A different type of bond occurs when two atoms share one or more electrons. Generally, the electrons are bound to the nucleus of one atom, but when two atoms are in close proximity some of each of their retinue of electrons can be shared between them. Essentially these moonlighting electrons enjoy the attraction of both nuclei and as a result bond the two atoms together. Molecular oxygen, or O_2, is an example of this type of 'covalent' bond. Molecules can bind together en masse in different combinations to assemble the material world we see around us. The exact mixture and arrangement of atoms, each with different numbers of protons, neutrons and electrons, determines the properties of these materials, and it's the electromagnetic force that holds everything together.

So where do electromagnetic *waves* come into this? Every charged particle generates a 'field' around it. This field is really just a way of describing the strength of the force around a charge and the effect it would have on other charges nearby. We usually sketch the field as a set of lines radially emanating from a charged particle like spokes, stretching away to (in principle) infinite distance. Near the particle, the density of the lines is high, and this means the field is strong. As we move away from the particle the lines are more spread out, meaning the field is getting weaker. Of course, in reality the lines don't really exist, they are simply a visualization of what we call in general a 'vector field'. They describe the path of a charged 'test' particle placed somewhere within the field. For example, an electron placed near a proton would feel the force and accelerate towards the proton along a field line.

In this basic picture we are considering stationary – static – charges and fields, hence the word 'electrostatic'. What happens when the charges move?

Imagine taking an electron and wiggling it about. Like a bug trapped on a pond, the agitated electron causes the electric field around it to ripple accordingly. The electric field is no longer static – it is moving. We are now talking about electrodynamics.

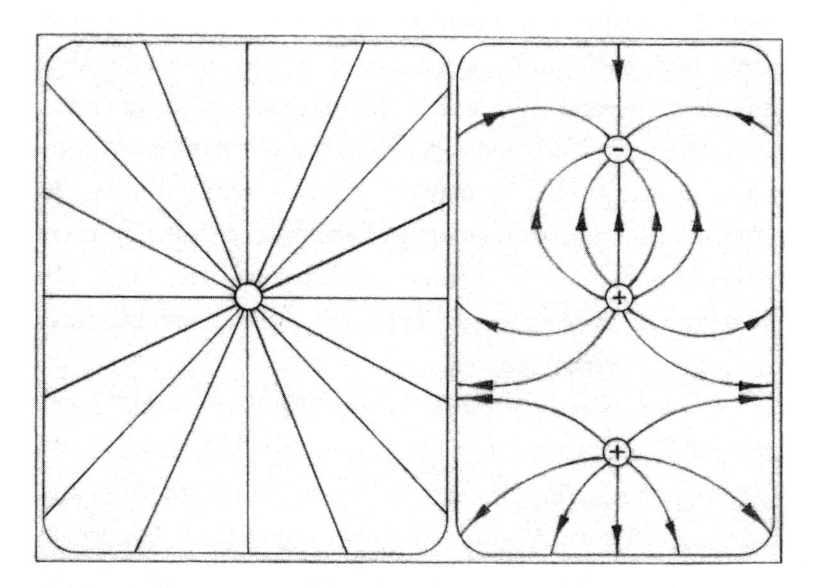

Field lines

We visualize the electric field around charged particles with 'field lines', describing the strength and direction of the electrostatic force felt by other charged particles within the field.

This is where the magnetism part of electromagnetism comes in, and it is key for the story of light.

James Clerk Maxwell (1831–1879) is considered the founder of what we now call the classical theory of electromagnetism. Maxwell, a Victorian physicist from Edinburgh, took forward the ideas of Michael Faraday (1791–1867), another pioneer of the field, and others to 'unify' electricity and magnetism, demonstrating how they are inexorably tied together. Maxwell showed that an electric field that is changing with time will induce a magnetic field, and vice versa. He also showed how the behaviour of those fields relates to charged particles.

Maxwell's achievement can be written in four elegant equations – Maxwell's Equations – that describe the 'classical' properties of the magnetic and electric fields and how they are linked. There's no need to write them down here, but the bottom line is this: Maxwell's Equations express something fundamental about the Universe. In his mathematical expression of electromagnetism, Maxwell showed that oscillations in an electric field generate an associated oscillating magnetic field, and these oscillations propagate away from their origin through 'free space' – that is, through the empty Universe – like a wave. This wave transports energy through space, which we call electromagnetic radiation. This is light.

How fast do these electromagnetic waves propagate? Using Maxwell's Equations it is possible to derive a 'wave' equation, which is a general expression for describing the properties of a wave – be it a water wave or an electromagnetic wave – in time and space. In the wave equation derived from Maxwell's work there appears a numerical constant, given the symbol c. It stands for *celeritas* and refers to the speed of propagation of electromagnetic waves through free space. c is the speed of light through a vacuum, also known as the speed limit of the Universe: two hundred and ninety-nine million, seven hundred and ninety-two thousand, four hundred and fifty-eight metres per second (299,792,458 m/s).

Just like waves on water, there is a simple way to characterize electromagnetic waves: by their wavelength or, equivalently, by their frequency. Both are related to the energy of the wave. The wavelength is simply the physical distance between two consecutive wave peaks. The frequency is the rate at which successive wave peaks pass some fixed reference point, and we measure that in units of wave cycles per second, also known as hertz (Hz). So, if the speed of a wave is constant, as it is with an electromagnetic

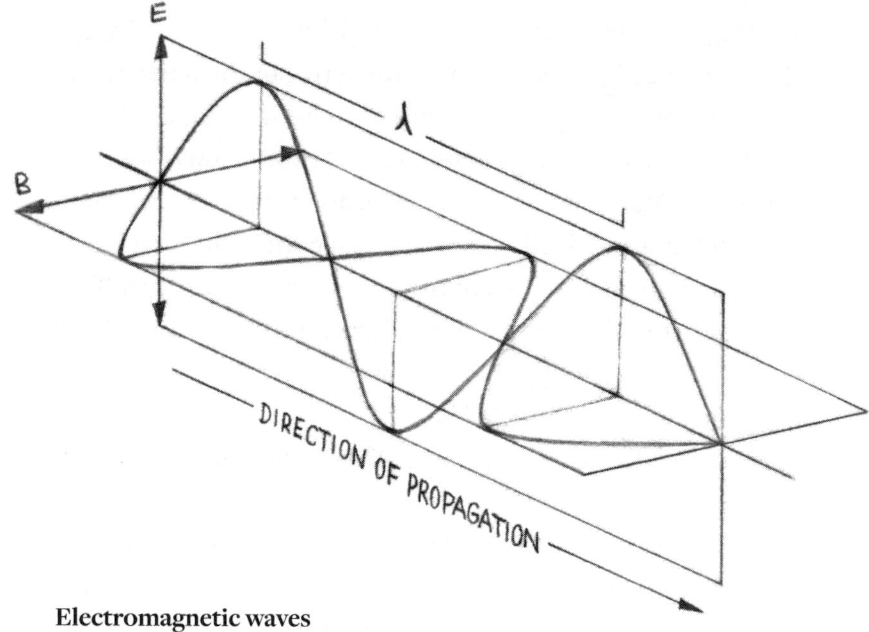

Electromagnetic waves

Oscillations in an electric field (E) will cause an associated oscillating magnetic field (B) perpendicular to it. These coupled oscillations propagate through space as an electromagnetic wave – light. We characterize the wave by the distance between peaks (the wavelength) or the rate of the oscillation of the fields (the frequency).

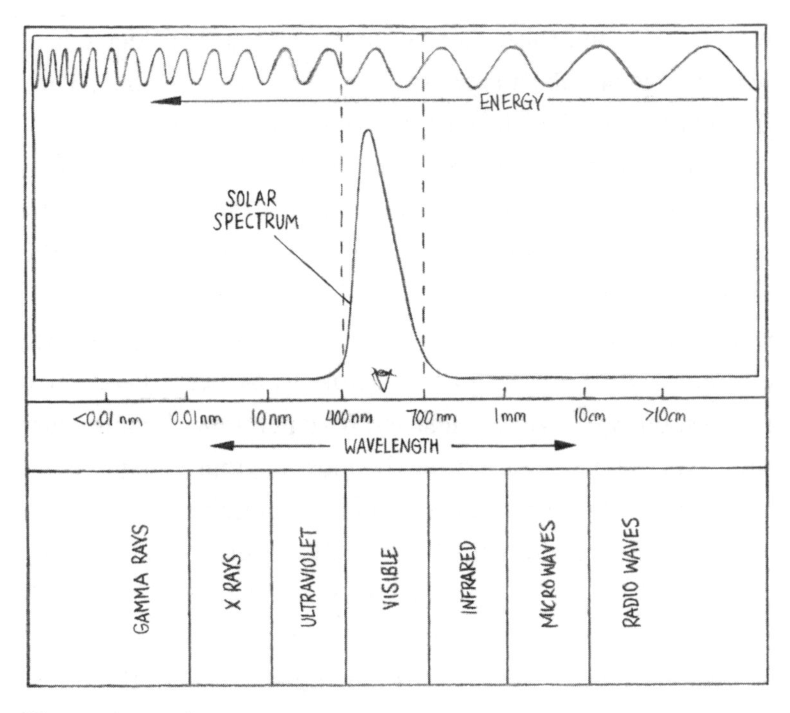

Electromagnetic spectrum

Electromagnetic waves, or photons, come in a wide and continuous range of energies, characterized by their wavelength (or, equivalently, frequency). The range that humans can see roughly matches the peak range of energies of light emitted by the Sun.

wave, then you can see that a longer wavelength corresponds to a lower frequency, and vice versa.

As we increase the energy of a wave, we shorten its wavelength and increase its frequency. Imagine taking the end of a long rope and wiggling it up and down at different rates: you will create 'waves' going down the rope with different wavelengths. Your arm is like a source of waves: wiggle your arm more vigorously and you'll increase the frequency of rope-waves. In nature we encounter electromagnetic waves – radiation – with a very wide range of energies depending on the source. In fact, we refer to an unbroken 'continuum' of energies of electromagnetic radiation, like a radio transmitter you could dial to any frequency. We call it the electromagnetic spectrum.

Although this is a smooth continuum, the range of energies is so large that it has been convenient to split the electromagnetic spectrum into chunks, and we give each chunk a label. At the lowest energies we have radio waves, which are electromagnetic waves with wavelengths of a few centimetres up to a kilometre or more. Radio waves can occur naturally, particularly in astrophysics, but we have also learned to generate and manipulate these waves for practical uses, the most obvious being in communication. Oscillating electric and magnetic fields will cause charged particles within the fields to also oscillate, and so a radio wave passing through an aerial will make the electrons within it respond. This response generates an electric current that can be measured by a receiver. By encoding information in the transmission of radio waves, we can communicate things like radio and television programmes over long distances.

Moving to a higher energy we come to the microwaves, which have wavelengths of millimetres to a few centimetres. These waves can cause water molecules in food to become agitated. The reason is, again, due to the response of charged particles to electromagnetic fields. Water molecules, made of one oxygen atom and two hydrogen atoms, are called 'polar' molecules because one end of the molecule is slightly positively charged and the other end is slightly negatively charged. This imbalance, called a 'dipole', means that when they are subjected to an oscillating electromagnetic field of just the right frequency, the water molecule will rotate. This molecular rotation is a form of energy, which is dissipated through the rest of the food as thermal energy, cooking it.

Continuing our journey up the electromagnetic spectrum, after microwaves we come to infrared radiation, with wavelengths of about one millimetre down to a few thousandths of a millimetre. This is quite a wide range, so infrared radiation is split into three sectors: far-infrared, mid-infrared and near-infrared. Those prefixes refer to the difference in energy from the visible light part of the spectrum, with near-infrared light lying just beyond the reddest light we can see.

It is useful to split the infrared part of the spectrum in this way because there are rather a wide variety of sources of infrared radiation in astrophysics. Any object with a temperature greater than a few tens of degrees above absolute zero will emit infrared radiation, and we often call this 'thermal' infrared emission. Once again, the reason is down to the motion of the particles within a heated body. To have a 'temperature' implies that the atoms and molecules are agitated: jostling around with thermal energy. At a temperature of absolute zero, the particles are stationary, but turn up the heat and they start to move, at least within the shackles of their bonds. The higher the temperature, the more violent the jostling. All these moving charges cause oscillations in their electric fields, generating electromagnetic waves that propagate into space, carrying away the thermal energy. This is the infrared glow. Low-temperature objects a few tens of degrees above absolute zero will emit long-wavelength, far-infrared light, and as the temperature increases the emission moves to shorter wavelengths, through the mid-infrared and then into the near-infrared.

Then we come to the visible part of the electromagnetic spectrum. Humans, and of course other animals, have evolved to see electromagnetic radiation with a wavelength range of about four hundred to seven hundred billionths of a metre (one billionth of a metre is called a nanometre). Think about what this means for a moment. Inside your eyes are cells that can actually respond to a small set of the electromagnetic waves that criss-cross space. More than this, the cells can transfer this response to your brain, which can decode the stimulus into meaningful information: images.

Beyond the bluest part of the visible spectrum we have the higher energy ultraviolet radiation, with wavelengths of tens to a few hundred nanometres. Similar to the infrared bands, this is split into 'near' and 'far' ultraviolet, with near ultraviolet light just beyond the bluest light we can see. We have all heard about the damaging effects of ultraviolet radiation from the Sun. It is damaging because of the energy of the wave; when it smashes into something – say some biological tissue – the energy being transported through space can be transferred to cellular matter. Sometimes this is to destructive effect, giving you sunburn, or worse, damaging molecules of DNA, which could go on to cause cellular mutation.

Ramping up the energy even higher, ultraviolet radiation gives way to X-ray radiation, and beyond this we encounter the gamma rays, with wavelengths of one hundredth of a nanometre and lower. This is where things get really dangerous. X-rays are useful because they can penetrate soft materials easily. Ordinarily you can't see into your hand, because visible light is absorbed by and reflected from the surface of your skin. X-rays, however, go straight through. We can use this to take images of the insides of our bodies, typically revealing the bones, which are more opaque to X-rays than skin and muscle. But this penetration can be a problem: like ultraviolet light, the high-energy X-rays can also cause damage to cells inside our bodies when their energy is deposited. Sometimes this might actually be desirable – we can use focused high-energy electromagnetic radiation to try to kill cancerous cells, for example.

On Earth, gamma rays are typically associated with radioactive elements and are simply a more extreme cousin of the X-rays. Exposure to such radiation can cause severe damage, and so sources of gamma rays must be heavily shielded by thick layers of dense material, such as lead, that can intercept as many of the rays as possible before they can do any harm.

We can imagine all these waves around us, flying this way and that in three dimensions, an ocean of oscillating electric and magnetic fields washing over us. These waves have different sources, which determine their energy. Most of them pass us by, or go straight through us, unnoticed. Some we can sense. We have harnessed some of them, transmitting, manipulating and detecting them for our benefit, be it transmitting tonight's episode of *The X Factor* or a session of radiotherapy. The point is, the waves are real; they are all over the place, travelling through space. That's what light – electromagnetic radiation – actually is. But there's a complication. There's another way of thinking about light: not as a wave but as a *particle*.

In the early twentieth century there was a revolution in our understanding of the natural world at its smallest level. You've probably heard of it: quantum mechanics. Now, quantum mechanics is a deep and complex subject, and unfortunately we don't have the time to delve too deeply into its wonders. Let's leave that for another time. The important thing is that quantum mechanics provides us with a framework for describing light and its interaction with matter that runs to a far deeper level than the 'classical' picture of electromagnetic waves.

The central principle of quantum mechanics is easy to grasp: energy, including electromagnetic radiation, comes in discrete chunks, called quanta. It is these quanta that can be thought of as particles. The idea that light is comprised of a flow of particles is not new. In the seventeenth century, Isaac Newton proposed that light was made up of 'corpuscles', or infinitesimal particles, and of course Newton was no crank. Unfortunately, this early particle model could not explain some of the observed behaviours of light, such as the patterns that are produced when light shines through small apertures. Eventually the wave model of light, championed by Newton's rivals and contemporaries, including Robert Hooke and Christiaan Huygens, became the accepted model and dominated our 'classical' thinking for centuries. It was not until the classical theory of electromagnetism itself started throwing up problems that we realized that the wave model couldn't be the end of the story.

When observations of natural phenomena cannot be explained by current theory, we have an opportunity to refine our understanding of how the world works. In science we develop theories that make predictions to be compared with observations. If these predictions don't match the observations, then the theory is refined or rejected. Most of the time these refinements are subtle, but occasionally they can be revolutionary. This is what happened at the start of the twentieth century.

One of the problems with the classical picture became known as the 'ultraviolet catastrophe'. It sounds a bit more dramatic than it was. The name refers to an issue that arose when scientists started to think about the radiation that would be emitted by a hypothetical heated cavity – a sealed box with a tiny hole drilled in it. Think of it as a kind of oven. A special property of this cavity is that its walls absorb and then re-emit all the radiation hitting them, reaching what we call 'thermal equilibrium' with the electromagnetic radiation. We call this a 'blackbody'.

Electromagnetic waves continually bounce around the walls, being absorbed and re-emitted until eventually some of them escape through the hole, to be observed.

In the classical theory, when it is in thermal equilibrium, the cavity is filled with a series of 'standing' electromagnetic waves, which are waves in which the amplitude of the field is changing, but the positions of the peaks and troughs are fixed in space. We often imagine these waves as a set of strings inside the cavity that are attached at fixed points on opposite sides of the walls. We can vibrate these strings like a guitar, and they oscillate with different frequencies. The different frequencies are called 'modes' of oscillation. Now, the classical theory states that the average amount of energy associated with each mode is proportional to the temperature of the system, and that you can, in principle, fit an infinite number of modes of increasing frequency inside the cavity.

If the total amount of energy in the cavity was divided between the different modes of oscillation of the electromagnetic field in the way the classical theory predicted, then the spectrum of the light emerging from the hole would be expected to diverge towards higher frequencies. In other words, the theory predicted that the intensity of light being emitted by the cavity should increase with increasing frequency, such that the total integrated emission of radiation from the hole becomes infinite. This clearly wasn't the case in practice.

Although the classical theory did a reasonable job of modelling the spectrum of electromagnetic radiation escaping from the cavity at low frequencies, it went drastically wrong once you reached the frequencies corresponding to ultraviolet light. Actually, the spectrum of light coming from the cavity follows a rather particular distribution: the intensity rises to a

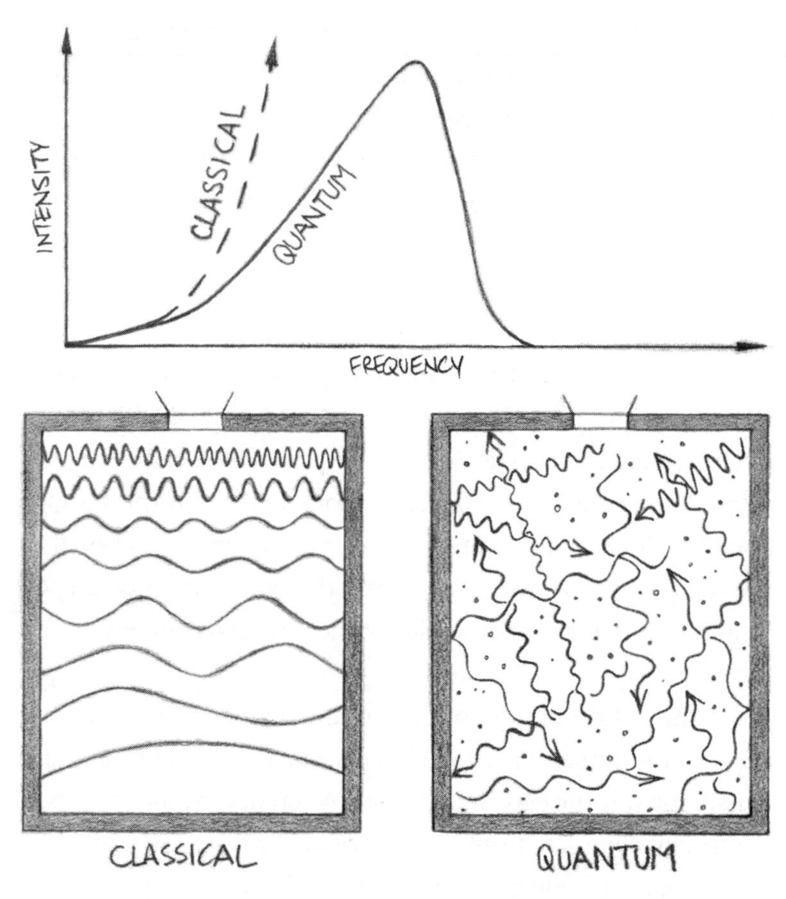

Cavity radiation
A comparison of the classical and quantum views of electromagnetic radiation filling a hypothetical heated 'blackbody' cavity in thermodynamic equilibrium.

peak at a particular frequency and drops away again at higher frequencies. The exact frequency of the peak intensity does depend on the temperature of the system, with hotter cavities emitting higher frequency (bluer) light, but the integrated, or total, emission is decidedly finite.

Max Planck (1858–1947), considered the father of quantum theory, proposed a solution right at the start of the twentieth century: instead of thinking about continuous waves, consider that electromagnetic energy is transported in discrete packets, called quanta. Planck addressed the problems arising from the classical picture using his quantum model, filling the hypothetical heated cavity with a 'gas' of particles, each a quantum unit of electromagnetic energy, rather than a set of standing waves. He derived a theoretical form for the intensity spectrum that matched observations beautifully. We now call those quanta of electromagnetic energy 'photons', and the functional form describing their distribution of energies in the heated cavity the 'Planck function'.

The discovery of the quantum nature of light and matter was a turning point in our understanding of the Universe. It not only gives us the atavistic satisfaction of knowing how things actually work, but in the true Baconian sense also gives us power, for it allows us to build technology. We can manufacture things like computers and flat-screen televisions, magnetic resonance

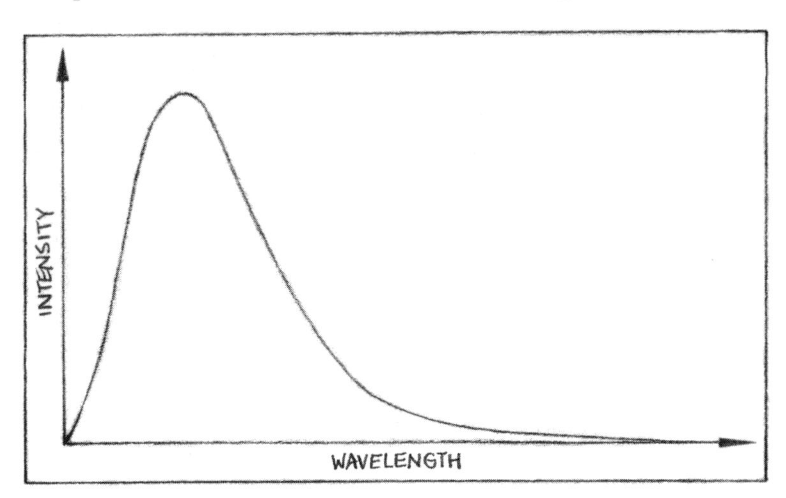

Blackbody spectrum

A blackbody spectrum, described by the Planck function. The position of the peak depends on the temperature of the body: hotter bodies peak at shorter (bluer) wavelengths and vice versa.

imaging (MRI) scanners and lasers because we understand how quantum physics works.

So do we need to throw away the notion of electromagnetic waves? Not at all. But here's where things get a little weird. Since energy can only come in discrete packets, and light is a form of electromagnetic energy, it means that at a quantum level the propagation of electromagnetic waves is more like the journey of single particles of energy: photons. However, we also know that light still obeys principles of wave physics, like refraction. So light can behave like a wave *and* a particle at the same time. This is called wave–particle duality. It is not just some theoretical quirk; the effect can be seen in experiment.

One of the most celebrated experiments – one performed by all students of physics – is called Young's Slits, after the scientist Thomas Young (1773–1829), who devised a version of the experiment in the early nineteenth century. Young's Slits is a simple experiment: the modern set-up consists of a pure light source, such as a laser emitting a single frequency of light, shining on an opaque sheet that has two very narrow slits in it. Each slit is no more than a fraction of a millimetre wide, and the slits are separated by about a millimetre. On the other side of the slits, some suitable distance away, is a projection screen. What we see projected on the screen is a series of bright and dark fringes.

In the classical wave picture, the laser is sending a series of plane waves towards the opaque sheet, like a set of parallel sea waves lapping up against a wall. At the positions of the slits, the waves can get through, but instead of progressing as a set of uninterrupted parallel plane waves, they fan out from each of the slits in an expanding semicircular pattern, as if the opening of each slit itself is a source of light. This effect is only noticeable if the widths of the slits are comparable to the wavelength of the waves passing through. You can see the same thing when sea waves enter the mouth of a narrow harbour, and in fact this bending of wavefronts happens for all types of wave passing through apertures or around objects where the width of the gap or size of the object is comparable to the wavelength of the wave. It is called diffraction and is another fundamental principle of wave physics.

It is what happens next that is key to the experiment. The semicircular waves emerging from the two slits will interact with each other because the wavefronts will start to cross each other's paths. In some places two peaks of the waves will align, and their amplitudes will increase accordingly. In other places a peak and a trough will coincide, cancelling each other out. We call this 'interference', and where those amplified and cancelled-out parts of the propagating wavefronts intercept the projection screen, we see bright and dark fringes, or an 'interference pattern'. This is all well and good, and easy to model and understand when we think of light as a wave. But what about the photon model, where light is a particle?

We can imagine the limiting case, in which we can turn down the original light source like a dripping tap so that it just fires single photons towards the screen. In the naive picture, some photons will hit the opaque part of the sheet and be absorbed, while others will happen to pass through one of the slits. In this case we might intuitively expect that because a photon either goes through a slit or it doesn't, eventually we would just see two bright bands appear on the projection screen, corresponding to the accumulation of photons that make it through. But this is not what is seen. The interference pattern of bright and dark bands still appears. This is because the photons still have the properties of waves, despite being discrete packets of energy.

The take-home point is this: in quantum mechanics, the actual location of a particle in space is no longer specific and absolute, so we can't really think of these quantum particles as well-defined, point-like entities flying around. Rather, until it is actually observed, like when it hits the screen, we can only describe a *probability* that a quantum particle – like a photon – exists at a particular location in space (and time, for that matter). We describe this probability distribution in space and time mathematically by the particle's 'wave function'. The wave function is a particular solution to a type of wave equation called the Schrödinger equation, named for physicist Erwin Schrödinger (1887–1961), another quantum pioneer. As we know, a wave equation is a mathematical formalism that describes the rules for the behaviour of waves, which could be water waves, acoustic waves or other types. It is the wave function that allows a quantum entity like a photon to behave both like a particle and like a wave. It is hard for us to imagine because in our macroscopic world we are used to things being 'definitely there', but on the quantum level, things are only 'probably' there.

This wave–particle duality applies to other quantum objects too. We have introduced electrons as charged particles, so you probably have in your mind a picture of a tiny ball. But an electron is very much a quantum entity, and it too has a wave function. It is not really in one specific place until it is measured, when we 'collapse' the wave function. So, if you repeat Young's Slits with electrons instead of photons, shooting them at the slits one at a time and then record the signal as they hit the projection screen beyond, you still observe an interference pattern – the accumulated electron signal emerges as a pattern of bright and dark bands. The wave functions of the electrons have constructively and destructively interfered between the

Young's Slits

A classic physics experiment involves shining light at two narrow slits set a little distance apart. If the width of each slit is comparable to the wavelength of the light, the waves passing through the slits emerge in a fan-like pattern; an effect called diffraction. The two sets of waves emerging from each slit 'interfere' with each other: two peaks will combine, and a peak and a trough will cancel out. This interference results in a series of bright and dark fringes on the final projection screen. The same pattern is observed when a series of single photons (or even electrons) are fired at the slits, demonstrating wave–particle duality.

slits and the screen. It's a beautiful demonstration of wave–particle duality.

Niels Bohr (1885–1962), another of the founders of the field, said that anyone who isn't shocked by quantum mechanics hasn't understood it. Indeed, there is a strangeness that comes with quantum physics that requires you to let go of your everyday intuition, and that's because we don't notice quantum effects in everyday life. But despite this strangeness, when I think about light I don't have the classical picture of a choppy sea of waves filling space, but something resembling more what Newton imagined. I think about streams of individual particles of light racing through space; little packets carrying different amounts of energy that are reflected or scattered or absorbed by matter along the way. Each has a journey.

Classical physics can get us so far, but without quantum mechanics – that deeper knowledge of nature – it is impossible to fully understand the interaction between light and matter at a fundamental level. The reflection of light from different objects, giving rise to different colours, the absorption of an ultraviolet ray by a cell in your skin, or the refraction of light as it passes through the lens in a pair of spectacles are all ultimately explained by the theory of quantum electrodynamics. Sometimes said to be the most perfect description of nature we have, quantum electrodynamics fully describes the interaction of photons with charged particles. In fact, the electromagnetic force itself can be understood in terms of the 'exchange' of photons between charged particles.

The basic rules that govern how photons and matter interact are the same for the light illuminating my living room and for the journey of every photon across the Universe. They describe every sunset and rainbow and moonlit night. They explain long shadows on a summer afternoon and the specular glint of sunlight off the hydrocarbon seas of Titan. They allow you to distinguish a Renoir from a Picasso.

Often our experience of light is mundane and tediously terrestrial: the bland illumination of fluorescent strips in an office block or the glow of the fridge door. But turn your gaze skywards and the story changes. You may see the crescent Moon, where the Sun casts shadows of the lunar mountains and crater lips, giving the terminator its jagged edge. You may see a planet reflecting sunlight across the Solar System, shining brightly in the evening sky. You may see thousands of stars pin-pricking the sky, and the ethereal band of the Milky Way, the starry plane of the galaxy in which we live. These sights have captivated humans for tens of thousands of years, but they are like the first few pages of a deeper story.

The story of the Universe is written in the light that fills it. Stranded on the surface of our tiny rock, we can look up and try to read that story. These are some of those tales: five astrophysical processes explained through the remarkable journey of light across space and time. Five photons.

TWO

OLD LIGHT

Look to the horizon. What is the furthest thing you can see? On Earth the extent of our gaze is limited by the curvature of the globe: tall ships appearing on the line between sea and sky reveal themselves mast and sail first before the hull comes into view. We can't see over the horizon, because light rays travel in straight lines.

What about the Universe itself? We can see, in principle, as far as the source of any photon that has taken less time than the age of the Universe to reach us. That sounds slightly convoluted, doesn't it? Perhaps more intuitively, think of it this way: light travels at a fixed speed, which means it takes a finite amount of time to reach us from wherever it was emitted, be it a light bulb in the same room or a distant star. Any photon that takes longer than the age of the Universe to reach us hasn't done so yet. We can't observe it. That sets the limit of the size of our *observable* Universe. The actual Universe as a whole could potentially be much larger of course.

The age of the Universe is around fourteen billion years old, so you might think that the horizon of the observable Universe is about fourteen billion light years away in every direction. It is actually much larger than that. The reason for this is that the Universe – space itself – has been expanding ever since it came into existence. Imagine a photon emitted by some hypothetical light source at the start of the Universe about fourteen billion years ago. The maximum flight time of that photon is, of course, set by the age of the Universe. In a static Universe this would indeed limit the distance travelled by the photon to about fourteen billion light years, but the separation between the source of that photon and us, the observers, has been increasing all the while during its flight because the Universe has been expanding. There is literally more space between us and the source of that photon now compared to when the light was first emitted. When you take this expansion into account, it turns out that the radius of the observable Universe is about *forty-five* billion light years.

As an astrophysicist who studies distant galaxies, I rarely think about photons having travelled great distances across space. What interests me is the length of the journey. Those photons have travelled across time as well as space. They come from the past.

Have you ever gazed at a sunset and watched the reddening solar disc as it plunges behind the horizon, and contemplated that you are looking back in time? The light you see from the Sun took about eight minutes to cross the Solar System before hitting the Earth's atmosphere. So your image of the Sun is about eight minutes old. In a stolen glance with a loved one across a room, you are seeing them as they were, momentarily, in the past. Light took a finite time to travel across the room from their eyes to yours.

Since cosmic distances are vast, the travel time for photons reaching us from distant astronomical sources is significant even considering the speed of light. The light from even the nearest star beyond the Sun takes a good four years to reach us. You can see here where we derive the astrophysical distance unit 'light year': it describes the distance travelled by a photon in one Earth year – equivalent to about ten thousand billion kilometres. Inevitably this means that when we actually 'see' an object – that is, when we observe and record its light – we see it as it was at the moment the light was emitted, not as it is right now.

We can turn this to our advantage, because it allows us to study the Universe as it was in the past simply by making very deep observations of the sky, detecting light from the most distant, faintest objects. The light reaching us from other galaxies can provide us with snapshots of the history of the observable Universe stretching billions of years into the past. By comparing the state of the Universe – such as measurements of the properties of galaxies – at a range of different distances, or rather, different light travel times, we can piece together how the Universe has evolved over cosmic history. Which raises the question, what is the oldest light we can see?

As we know, not only does light travel at a fixed speed, the Universe has a finite age. This places an obvious upper limit to the oldest light, since there are no photons from before the Universe existed. So can we see the light that was emitted right at the start of the Universe? Can we see the moment of creation? Frustratingly, the answer is no – or at least, not quite.

From an electromagnetic point of view, we don't have a clear view of the Universe before it was about 380,000 years old. This is due to the thermodynamic properties of the matter content of the cosmos and its interaction with the electromagnetic radiation that permeated it in the few hundred thousand millennia after it formed. Luckily, this phase of the Universe has imprinted an observational signature on the radiation from that time that we can now detect. This is, in itself, a key piece of evidence that our Universe originated in a 'Hot' Big Bang.

The Hot Big Bang describes the process by which all of space and time originated from a single point, when the entire contents of the observable Universe were compressed into an unimaginably dense volume, or 'singularity'. That singularity literally contained everything that now makes up you and me, and all the stars, and all the distant galaxies and everything in between. Quite simply, the whole shebang was all in the same spot, seething with the potential for the vast, unfathomed Universe that we find ourselves in today. It is remarkable to think that the matter in your body, in one form or another, was once in the same intimate cosmic volume as everything else in the entire Universe, a cosmic seed. Around fourteen billion years ago this seed rapidly expanded like an explosion. It is tempting to think of the explosion happening in some pre-existing volume, like a blank canvas, with matter and energy spilling out. But space and time itself were also created in this event, and the matter and energy content of the Universe was dispersed within this expanding framework, which continues to expand to this day. The 'explosion' happened everywhere.

One of the first pieces of evidence for an expanding Universe was noted by Edwin Hubble (1889–1953) in the 1920s, when he combined observations of stars in 'spiral nebulae' with Vesto Slipher's spectroscopic measurements of the apparent 'recession velocities' of the same objects. One of the types of stars Hubble observed are called Cepheid variables.

Cepheid variables are stars that increase and decrease in brightness in a periodic way. They physically expand and contract over days or weeks, allowing more or less light to escape as the stellar atmosphere becomes more or less transparent to photons over the cycle. Think of it like a mesh bag filled with sand: stretch the mesh and the holes get bigger, letting out more grains; when it shrinks back down, the flow of sand slows. Henrietta Swan Leavitt (1868–1921) had previously shown that the period of Cepheid variable stars' regular beat in brightness is strongly correlated with their intrinsic luminosity, a relationship that is known as 'Leavitt's Law'. This is important, because if you can measure the luminosity (think of it as the wattage of a bulb) of a light source independently of its observed brightness, then you have a good idea of how far away it is: the observed brightness should be proportional to the intrinsic luminosity, and inversely proportional to the square of the distance from the star. Cepheid variables were one of the key methods Hubble used to measure the distance to spiral nebulae, using Leavitt's Law.

Slipher (1875–1969) was a key figure in the discovery of the expanding Universe, but like Leavitt he unfortunately doesn't get the same credit as Hubble. Slipher had carefully observed the spectra of various spiral nebulae, allowing him to split the light according to its energy, like sunlight through a prism. Spectra reveal far more information than a rainbow of colours; they are a bit like a barcode, imprinted with bright spikes and dark troughs where photons with certain energies are being preferentially emitted or absorbed by particular elements. A gas cloud that contains oxygen atoms, for example, might be ionized by the radiation emitted by a nearby star. The oxygen ions can then give rise to a glow of radiation of a very specific frequency, appearing as an 'emission line' spike in the spectrum at that frequency. You have probably encountered this phenomenon without knowing it: a regular sodium-based street lamp has that characteristic yellow glow because it contains sodium ions emitting photons with a wavelength of about 590 nanometres. It is not a broad spectrum source. We know the value of the wavelength (or, if you prefer, frequency) of the light emitted by many different elements and their various ionized states, either from laboratory tests or from atomic theory. So when we measure the spectrum of a distant astronomical source we can tell which emission line comes from oxygen, and which comes from, say, neon, and so on. So how do we measure a 'recession velocity' from this information?

Recession velocity refers to the observation that spectral features (such as emission lines) observed in objects like the spiral nebulae Hubble was interested in, can appear shifted in wavelength from their expected 'laboratory' positions. It is like the barcode has been shifted backwards or forwards along the wavelength scale. We call these 'redshifts' if the wavelengths are longer than expected, or 'blueshifts' if the wavelengths are shorter than expected.

The cause of this effect is akin to the 'Doppler effect' in the pitch of a police siren as it approaches and then whizzes by you. The siren is emitting acoustic waves with some characteristic wavelength, and that determines the pitch. But the distance between the peaks and troughs of the sound wave as detected by *your* ears is being compressed (moving to a higher pitch) as the source of the waves moves towards you. The reason is that in

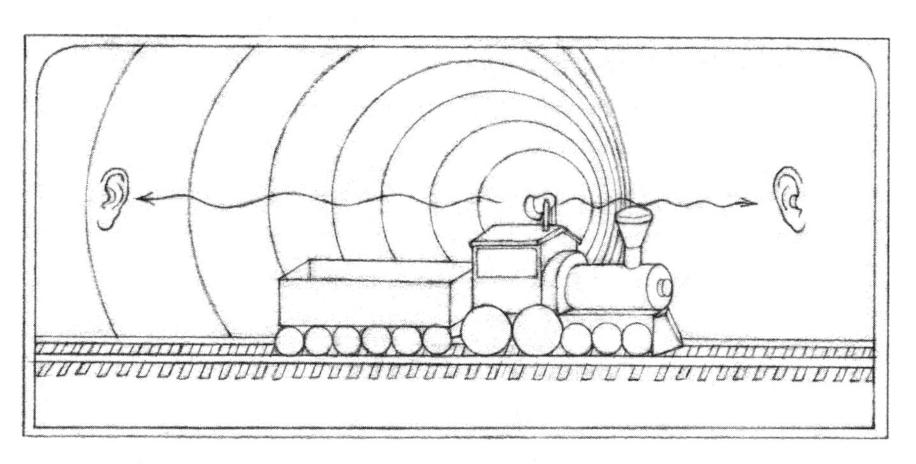

Doppler effect
The Doppler shift is the observation of a stretching or compression of the wavelength (equivalently a decrease or increase in frequency) of waves when the wave source is moving away from or towards the observer.

the time it takes for the siren to emit another wave, the car has moved towards you a little bit, catching up with the previous wavefront. The opposite is true as the car speeds away from you: the distance between peaks in the sound wave – the wavelength – is stretched, moving the sound to a lower pitch. To think of it another way, the time between successive wavefronts reaching your ears – the frequency – is getting lower, because the receding car is putting more distance between the waves than would be the case if the car was stationary, relative to you. The same thing happens with electromagnetic waves, except that the speeds involved tend to be much higher to have a noticeable effect. So blueshifts correspond to objects moving towards us, and redshifts correspond to objects moving away, or receding from us. The size of this shift can be converted into a velocity, relative to your frame of reference.

Hubble put the Cepheid variable and other distance measurements together with velocities measured from Slipher's spectra and showed that the spiral nebulae – what we now call galaxies – are generally redshifted, or receding away from us. What is more, the distant galaxies appear to be moving away faster than those nearby. The relationship between recession velocity and distance can be fit with a linear relation called Hubble's Law, characterized by a constant factor describing the recession velocity at a given distance. We call this the Hubble constant.

Although not explicitly described as such by Hubble in his 1929 paper presenting the result, this is now considered to be the first clear empirical evidence of the expansion of the Universe. Now, if one plays the cosmic tape backwards, reversing the expansion, there is the suggestion of a common spatial origin for all the galaxies: things that are moving away from each other must have once been closer together. Yet it was not until several decades later that the idea that the Universe originated in an all-encompassing cosmic explosion became an accepted model. More evidence was needed for such an outlandish claim. A key prediction was that if the Universe was much denser in the past because stuff was closer together, then it would have also been much hotter earlier in its history.

In the 1940s and '50s, cosmologists were theorizing about the possibility of detecting so-called 'relic' radiation from a Hot Big Bang. Back then, the idea that the Universe erupted suddenly and spontaneously from a singularity was new and radical and – more importantly – lacking firm empirical evidence, despite the work of Hubble and others decades before. The establishment scenario was that of the Steady State model, which posited that the Universe had always existed, and Hubble's observations were accounted for by postulating that new matter came into existence in the gaps that formed between galaxies as they moved apart. Actually, the term 'Big Bang' was coined around this time somewhat out of derision by the cosmologist Fred Hoyle (1915–2001), a key proponent of the Steady State model, but the name has stuck to this day.

The Achilles heel of the Steady State model was this: one of the main predictions of the Hot Big Bang model was that relic radiation from the first moments after the explosion (for want of a better word) might still be detectable today in the microwave part of the electromagnetic spectrum as an ambient 'background' all around us. If one could detect this cosmic background, it would be compelling evidence for the Big Bang and the death knell for the Steady State.

And we did detect it. The first observation of this background light is celebrated as something of a happy accident. In the mid-1960s Arno Penzias (1933–) and Robert Wilson (1936–), working at Bell Labs in Holmdel, New Jersey, were conducting observations with an instrument called a horn antenna. It looks like it sounds: in effect a large listening device, designed for detecting radio waves. Penzias and Wilson were researching something quite practical: the detection of radio waves bounced off 'echo balloons'. These balloons were launched to an altitude of about a thousand kilometres and could reflect radio waves. This was one of the first steps towards satellite communication; the idea was that one could bounce radio waves off the balloons and pick them up somewhere else, beyond the horizon.

To work properly, it was essential to eliminate all other sources of 'contaminating' signal by removing known radio emissions coming from regular broadcasts and also by cooling the instrument down to minimize ambient 'thermal' noise. But Penzias and Wilson always ended up with some low-level noise in their signal. It appeared in their observations independently of the time of day or season and was present no matter what direction the horn was pointing. Even the removal of 'white dielectric material' (pigeon droppings) from the horn didn't help; the mysterious signal remained. The frequency of that rogue signal was at about four gigahertz, close to the microwave part of the electromagnetic spectrum.

When Penzias and Wilson talked to the theorists at nearby Princeton University, they soon came to realize the implications of the background 'noise' they had stumbled upon. In 1965 they described their findings in a letter published in the *Astrophysical Journal* entitled 'A Measurement of Excess Antenna Temperature at 4,080 Mc/s [Megacycles per Second]'. The temperature Penzias and Wilson measured was 3.5 Kelvin (about –270°C), with about a one-degree margin of error. At the same time, the group of theorists, led by Robert Dicke (1916–1997), wrote a companion letter describing the theory behind the interpretation that the excess antenna temperature Penzias and Wilson had detected was actually the cosmic microwave background, the relic radiation from the Big Bang. This was a seminal moment in the history of cosmology, and Penzias and Wilson won the Nobel Prize in Physics in 1978.

Like most sources of radiation in astrophysics, the photons that make up the cosmic microwave background (which we shorten to CMB) are not all of the same energy, but have a characteristic distribution, or spectrum of energies. When sunlight is dispersed into a rainbow, the range of colours you see betrays the distribution of energies of the photons, because the energy of a photon is directly proportional to its frequency. Just as Slipher measured the spectra of galaxies in the visible part of the electromagnetic spectrum, we can build detectors that operate in other parts of the spectrum. That makes it possible to measure radio spectra or X-ray spectra or infrared spectra and so on, depending on the source of light.

When the spectrum of the CMB was first measured properly by an instrument called the Far Infrared Absolute Spectrophotometer on the Cosmic Background Explorer (COBE; pronounced 'co-bee') satellite, it was found to follow an almost perfect 'blackbody' curve: a slightly lopsided bell-like shape with a characteristic peak in frequency. We have already encountered this curve: it is mathematically described by the Planck function.

Conceptually, a blackbody is an object that absorbs all electromagnetic energy falling on it, like the heated cavity we met before. An object, or set of particles, that is in thermodynamic equilibrium – meaning that the thermal energy, or temperature, has reached a steady state everywhere – will re-emit radiation with the characteristic blackbody spectrum. The frequency, or photon energy, where the spectrum peaks is determined only by the temperature of the object: increasing the temperature shifts the peak to higher frequencies and decreasing the temperature shifts it to lower frequencies. Think of a glowing metal rod: it will first glow dull red, through brilliant orange and yellow until it turns a blue-white when it is really hot. The colour of the light emitted by the heated rod is related to the internal energy of the particles within it: they get agitated and release photons of increasing frequency as more thermal energy is introduced into the system. So, measuring the spectrum of an object that is behaving like a blackbody and finding the frequency at which the intensity of the radiation peaks allows us to measure the temperature of that object very accurately. That works even when the object is the entire Universe.

There are few objects in nature that behave as pure blackbodies. Sure, there are objects that are blackbody-like, but their spectra generally deviate slightly from the ideal theoretical curve. But the CMB turns out to be the most perfect natural example of a source of blackbody radiation known, and its spectrum matches the Planck function almost exactly. The CMB spectrum peaks at a frequency of about 160 gigahertz. If we turn that into a temperature we get a figure of a little over 2.7 Kelvin, which is within the uncertainty budget of Penzias and Wilson's original measurement. You can think of it as the ambient temperature of the Universe today.

At just under three degrees above absolute zero, this perhaps seems very cold. This is because what we measure today is not the temperature of the Universe at the time the CMB was actually emitted. Rather, we are detecting the relic radiation background after nearly fourteen billion years of cosmic cooling. The Universe is now just a glowing ember of its creation fire.

Immediately after the Big Bang, and for the following few hundred millennia, the Universe was what can only be described as a hot mess. For hundreds of thousands of years, despite constant expansion, the normal matter content of the Universe was completely ionized: electrons were not bound to protons, because their thermal energy was much higher than their electrostatic 'binding energy'. This meant that the Universe was permeated by a dense soup of protons and electrons, freely moving about. In among the protons and electrons were photons. Electromagnetic radiation. Light. The photons are the fire of the explosion.

Now, we know that photons are another way of thinking about electromagnetic waves, and we have learned that electromagnetic waves interact with charged particles. In the dense early Universe, the average distance between electrons, protons and photons was pretty small, so any photon zipping along soon encountered an electron. During these encounters the photon and electron undergo an electromagnetic interaction, and this results in the photon being 'scattered' in a random direction, only to scatter off another electron, and another, and another, and so on. Think of entering a crowded pub and trying to walk from the door to the bar. You'd like to take a straight-line path, but it's difficult to do because you are being constantly jostled. Your path becomes jagged. That's what it was like for photons in the early Universe. It's convenient to think of photons scattering with electrons as if they are tiny ballistic particles, with the photon travelling along, 'colliding', and then shooting off in another direction – but what is actually happening?

Remember that photons are a quantum manifestation of an oscillating electromagnetic field. Electrons, as charged particles, respond to that field: it causes them to oscillate too. In turn, the electron produces a photon with a frequency determined by the period of the oscillation. This photon is emitted in a random direction. So, an incoming photon interacts electromagnetically with the electron and is fired off in a new direction (we'll leave the question of whether it is the 'same' photon for another day). When the energy of the photon before and after scattering is the same, it is called Thomson scattering, after the physicist J. J. Thomson (1856–1940). It is possible for the scattering process to exchange energy between the photon and electron, and this is called Compton scattering, after the physicist Arthur Compton (1892–1962).

We can think of the average distance a photon travels before each scattering event as a way of characterizing the propagation of the photons. In physics we refer to the 'mean free path' – the average distance a photon can travel in free space before interacting with matter. If the mean free path is small, then the medium is opaque: the photon does not have a clear run between its point of origin and a distant observer. With each scattering event the photon loses the 'memory' of the previous part of its journey. An analogous situation is watching a friend walk towards you through a bank of fog. While they are deep within the fog bank you cannot see them at all because the photons reflecting off their body are readily scattered and absorbed by myriad tiny airborne drops of water. The larger the 'column' of fog between you and your friend, the more of those drops the photons have to contend with before they reach you.

While the column density is high – that is, while there is lots of fog in between – the probability of any one photon emerging from the fog to your eyes without interacting with a water drop is vanishingly small. We say that the 'optical depth' between you and your friend is high. As they walk closer, the column density of fog drops and so does the optical depth; the photons have fewer water drops to contend with before getting to your eyes. Some of them do make it right through in a direct path, and you start to recognize your friend. The probability of this happening gets higher and higher as your friend gets closer to you: more and more photons escape on a direct path, reflecting off their body and straight into your eyes. Eventually you can clearly see your friend emerging from the mist.

The fact that we can see stars and distant galaxies at all relies on the fact that the Universe is, on the whole, transparent to light. There's little 'fog' in between. That is to say, a photon from some far-flung star can travel pretty much unimpeded all the way to our eyes or digital detectors, 'free streaming' across space for perhaps billions of years. But in the very early Universe the photons were constantly scattering in a fog of free electrons. The optical depth of the Universe was extremely high; it was *not* transparent to these photons. The consequence for us is that we cannot see beyond this point, at least not with electromagnetic waves. It is the fog that blocks our view of the primordial Universe.

About 380,000 years after the Big Bang something changed. As the Universe expanded, it cooled, and indeed it continues to cool to this day. This means that the thermal energy of the protons and electrons dropped, and at some point it reduced sufficiently for electrons to be captured by the protons. Electrons were no longer free, but bound into atoms for the first time. The name of the simplest and most common element in the Universe is hydrogen: a single proton and electron bound together to form a neutral atom. There were some other atoms formed at the same time, such as deuterium, helium, lithium and beryllium – but hydrogen dominates the elemental budget.

So, all of a sudden (in cosmological terms) the Universe underwent a shift from being completely ionized to neutral.

Those photons were still zipping about but, with all the free electrons now bound into atoms, they were no longer scattered. Photons could now free stream, pretty much unmolested. We call this episode the Epoch of Recombination, describing the time when electrons 'recombined' with protons to form neutral atoms. Arguably this is a poor name for it, since the electrons were never combined with the protons in the first place, but astrophysicists have a rich history of poorly chosen nomenclature.

With the mean free path becoming effectively infinite, the primordial photons flew across space in whatever direction they were last going, as all the while the Universe expanded and evolved around them. The first stars in the first galaxies started forming a few hundred million years after recombination, assembling along great filaments and clusters. After about eight billion years or so, our own Solar System and Earth formed. All the while the photons that were 'released' at recombination were still racing through the Universe. Five billion years later, humans evolved, and eventually started to ask questions about the world around us. We figured out the rules of light and how to build telescopes and instruments to detect the photons raining down from the Universe beyond the bounds of the sky. Very recently, we began to capture some of those ancient photons that set off on their journey a little over thirteen billion years ago. They are now the cosmic microwave background.

If those photons started out with quite high energy in a hot Universe, why do we now detect them in the microwave part of the electromagnetic spectrum, at fairly low energy? Why has the fire faded? Imagine a photon flying across the Universe: throughout the journey, the Universe is expanding. It is not expanding *into* anything, but rather the scale of space itself is expanding, as if marks on the tape measure are getting further apart. It means

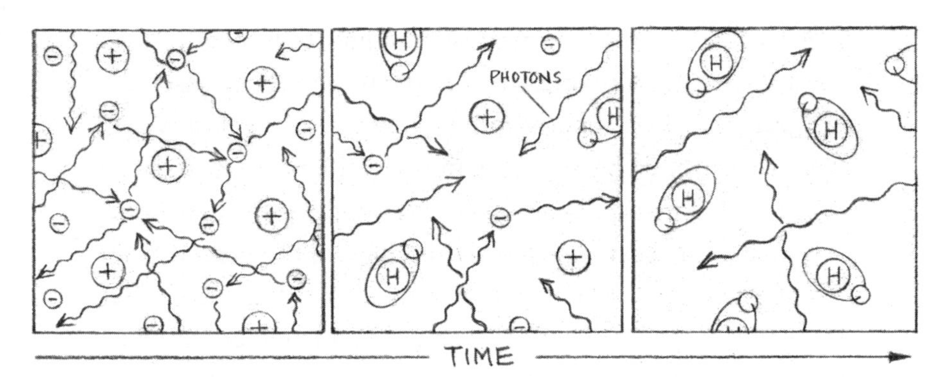

Recombination
Nearly 400,000 years after the Big Bang the Universe had expanded and cooled sufficiently for free electrons to combine with protons to form neutral hydrogen. Photons that were previously trapped within the hot plasma – because of their constant scattering interactions with electrons – were released. They are now seen as the cosmic microwave background.

that distant galaxies, and the origin of the cosmic microwave background itself, are receding away from us. Just as Slipher's spectroscopy of galaxies showed, this recession introduces a redshift of the light we observe from distant sources that can also be thought of in terms of the Doppler shift. A shift towards lower energies of electromagnetic radiation.

Imagine an observer located in every galaxy in the Universe. For now, let us ignore any local gravitational motion that would cause nearby galaxies to be accelerated towards each other, as is the case in the real Universe. All our observers are simply moving away from one another with the cosmic expansion, and an observer in one galaxy measuring the spectrum of a neighbouring galaxy would see a redshift consistent with the Doppler shift due to that galaxy's recession velocity, relative to them.

Now let's imagine we could communicate with our nearest neighbouring observers, and they could communicate with their neighbours, and so on, all across time and space. Let's also say that the means of communication is simply to rebroadcast the spectrum of the CMB as measured by our neighbouring observers. Think of this like a message being passed from observer to observer down a chain. Just as in the real Universe, the finite speed of light means that it takes some time for the message to carry down the chain. Furthermore, since all the observers are moving apart from each other due to cosmic expansion, each rebroadcast spectrum will be subject to a Doppler shift as it passes down the chain. So each time the spectrum is rebroadcast, it picks up a bit of redshift.

Rather than having observers in every galaxy, we could imagine a Universe filled with observers, each separated by a very small distance. In this limiting case the Doppler shifts between neighbouring observers becomes tiny, or 'infinitesimal'. Still, all these little redshifts across the expanding Universe accumulate to a final 'cosmological' redshift. So by the time we receive the spectrum of the CMB as originally measured by the most distant observer from us, it has been redshifted into the microwave part of the electromagnetic spectrum. Those photons have lost energy and now describe a much colder blackbody. The Universe has cooled in the time it took for the message to get down the chain.

Sometimes cosmological redshift is simply explained in terms of the electromagnetic waves being stretched as they traverse expanding space. This is not a bad description, but it gives the impression that all of space is expanding all the time, even on local scales, like the size of this book, and that's not correct. In fact, cosmic expansion can only be considered as a global property of the Universe. The picture of an accumulation of infinitesimal Doppler shifts between observers spread across the Universe is more appropriate, but note that it is not correct to think of the cosmological redshift due to cosmic expansion as equivalent to one single, global Doppler shift of the light between a distant source and observer.

The expansion of the Universe and its effect on the electromagnetic radiation travelling through it is why the peak of the background relic radiation has reduced in energy to microwave frequencies by the present day. Some hypothetical civilization that existed billions of years ago in what would have been a smaller, denser, hotter Universe would have also measured a relic background with a blackbody spectrum, but they would have seen it peak at a higher frequency than we measure today. Similarly, a civilization observing the Universe billions of years from now will measure an even colder relic radiation, with a blackbody peak that has redshifted even further, into the radio bands of the electromagnetic spectrum.

At the Epoch of Recombination, we can think of the CMB photons being suddenly 'released' after being trapped in the hot plasma. From then on, they have enjoyed a nearly unimpeded journey across the Universe. That means when we detect these photons and map them, we can visualize the environment from which they were emitted. And since the photons were released very quickly, and almost simultaneously everywhere in the Universe, it appears to us as if they were emitted from the inside of a surface all around us. Indeed, we call the origin of the CMB the Surface of Last Scattering, and it represents the most distant, and oldest, light we can see. If we measure the background radiation along every line of sight from the Earth, we can map this surface. Now, here's where things start getting interesting.

We have learned that the temperature of the CMB is about 2.7 degrees Kelvin, and that holds no matter what direction you look in. Measure the temperature along one line of sight and then turn around and do it in the opposite direction and you get almost exactly the same answer. When this observation was made, it backed up a central tenet of cosmology called the Cosmological Principle: that the Universe is isotropic – that is, statistically identical in all directions. No preferred direction. No bright centre. Nowhere particularly special to look.

But I said the temperature is *almost* the same everywhere you look. It turns out that the temperature of the CMB does fluctuate from place to place, but by tiny amounts. The temperature variations across the sky are typically around one part in one hundred thousand, or about a few tens of millionths of a degree. These fluctuations are called anisotropies, and encoded within them are clues about the very nature of the Universe and the origin of all the structure we see around us today.

The first 'proper' maps of the CMB were made in the early 1990s by the COBE satellite. COBE had a simple mission, to explore the microwave background, and one of the ways it did this was through an instrument called the Differential Microwave Radiometer. Its job was to measure anisotropy in the CMB temperature. Prior to COBE, anisotropies were predicted to exist, but had not yet been measured; it is no easy task. A key problem is the fact that we live in a galaxy that is itself bright with radiation in the same part of the electromagnetic spectrum as the CMB, mainly from the thermal emission of interstellar dust that is heated by starlight. Not only that, but the motions of Earth around the Sun, the orbit of the Solar System in the Milky Way, and even the movement of the Milky Way through space cause large-scale distortions in the observed temperature of the CMB: we measure it to be hotter in the direction of motion.

Any map of the CMB must take into account the galactic 'foreground' emission and any anisotropy introduced by our motion through space. Luckily there are very clever ways to remove most of these effects, developed over many years of observational experience, that allow us to peel back the various foregrounds and compensate for the distortions. This reveals the 'primary' anisotropies in exquisite detail.

COBE produced the first map of the temperature anisotropy of the CMB, but the telescope had quite low angular resolution and so it could only map differences in the CMB temperature on angular scales of about ten degrees or more. For comparison, the size of the full Moon in the sky is about half a degree across. Nevertheless, this was a tremendous achievement in the field of observational cosmology, for COBE had mapped the subtle differences in the temperature of the CMB that are a snapshot of the thermodynamic and structural properties of the Universe as it was just a few hundred thousand years after the Big Bang. So important was this breakthrough that George Smoot and John Mather, the two astronomers who led the project, won the Nobel Prize in Physics in 2006. The Nobel committee described COBE as the mission that was the 'starting point for cosmology as a precision science'.

And indeed it was. Since COBE, several more missions have mapped the CMB with ever-increasing sensitivity and resolution, both in space and from the ground. In the 2000s the Wilkinson Microwave Anisotropy Probe (WMAP) produced an updated map of the temperature anisotropies, revealing a finer and more sensitive view than COBE could. More recently, in the early 2010s, the Planck satellite produced an even sharper map.

There have been several ground-based observations of the CMB too, with two notable experiments of recent years being the Atacama Cosmology Telescope in the arid Chilean desert and the South Pole Telescope in Antarctica. These experiments have to be located in very dry parts of the world (although icy,

Antarctica is also extremely dry, so an excellent astronomical site) because water vapour in the atmosphere has the rather inconvenient property of absorbing photons in the far-infrared and microwave part of the spectrum. However, on the ground one can build much larger telescopes than can be flown in space, and with a bigger telescope you can make maps with finer detail and with better sensitivity. This allows us to measure the temperature anisotropies on even smaller scales.

Mapping the anisotropies is all well and good, but *why* is the background ever so slightly hotter in some directions and colder in others? The answer lies in the density of matter at the time of recombination. The distribution of matter in the Universe has never been totally homogeneous. Indeed, it is rather clumpy today. At its extreme, we see great clusters of galaxies and huge voids of empty space, and we observe galaxies strung along huge filaments that lace the Universe in a cosmic web. The origin of all this structure can be traced back to the hot, dense soup that existed at the time of recombination.

Although the Universe was filled with a plasma of protons and electrons, intermingled with particles of dark matter, it was not totally smooth. In some places the density of matter was a bit higher than average and in other places the density was a bit lower than average. These density fluctuations are thought to find their origin even further back in cosmic history, right to the singularity itself. Random density perturbations at the quantum level are predicted to have been amplified during a period of rapid 'inflation' at the Big Bang. Gravity has the effect of continuing to amplify the density perturbations, such that dense regions get denser by attracting more matter, which continue to grow, attracting more matter, and so on. The pattern of these 'seed' regions in three-dimensional space, and their evolution over time, is fundamental to our understanding of galaxy formation, since galaxies tend to form where the density of matter is high.

Close to the Epoch of Recombination, matter was compressed in environments that were slightly denser than the average. Locally, this made the plasma slightly hotter. In the corresponding environments that were of slightly lower than average density, the plasma was more rarefied, and cooler. The photons, trapped in the plasma by their constant scattering, were in thermal equilibrium with the electrons, meaning that the energy distribution of the photons was linked to the thermal energy of the electrons. So in the denser regions, where the thermal energy of the electrons was higher, the distribution of photon energies was pushed towards higher energies. Conversely, the distribution was pushed to lower energies in the low-density regions.

When recombination occurred, scattering stopped and the photons escaped from the plasma. Suddenly released, they carried away whatever energy they had at their last scattering event, across the Universe. The photons escaping from dense regions left a hot spot on the Surface of Last Scattering because their distribution of energies – encoded in the blackbody spectrum – peaks at slightly higher frequency than average. Those escaping lower-density regions left a cold spot because their blackbody spectrum peaks at a slightly lower frequency. The distribution of the density fluctuations in the primordial plasma at the time of recombination is then forever encoded in the distribution of energies of CMB photons. Over thirteen billion years later, we measure these as the one part in one hundred thousand temperature anisotropies, seen as a mottled pattern of hot and cold spots on the Surface of Last Scattering, frozen in time like a photograph.

One of the most elegant measurements in cosmology is the 'power spectrum' of the temperature anisotropies. The power spectrum is a way of statistically describing how the amplitude of temperature fluctuations is distributed on different scales on the sky. Since the original density fluctuations occurred on a range of physical scales, from broad swells to smaller ripples in the density of matter, we see temperature anisotropies on a range of angular scales projected on the sky. Measuring this distribution is a powerful probe of the nature of the early Universe and reveals some deep cosmological secrets.

The first thing you notice is that the power spectrum has a dominant peak: the strongest temperature fluctuations occur on a characteristic scale. This first big peak is followed by lots of smaller peaks, wiggling away with decreasing power as we look at smaller and smaller scales on the sky. We can translate angular sizes on the sky to a physical size, if we know the distance. Think of the apparent (angular) size of a penny held in front of you. It always has the same *physical* size, but its angular size gets smaller as you extend to arm's-length. The biggest peak in the power spectrum is at an angular scale of about one degree, which is about twice the diameter of the full Moon.

You may ask what is special about this particular scale? The answer lies in the properties of the plasma that permeated the Universe at the time of recombination. We know that the photons and electrons were interacting through a process called scattering. In turn, the negatively charged electrons were also interacting with the positively charged, but more massive, protons because of the 'Coulomb' force between them. Combined, the photon–electron scattering and electron–proton Coulomb interaction coupled together the photons, electrons and protons. Protons (and neutrons) are a class of particle called 'baryons', and it is these baryons that dominate the mass budget of 'normal matter' in the Universe. So the medium that filled the Universe at this early time is often referred to as the 'photon-baryon fluid'.

Mixed in with the photon-baryon fluid is the mysterious, non-baryonic dark matter that we think must be present in the Universe in vast quantities. The only way we know of that dark matter can interact with normal matter is through the gravitational force. It does not care about any of the electromagnetic interactions going on in the photon-baryon fluid, but it *does* care about the local density of matter, regardless of its form.

Gravity, of course, played an important role in shaping the landscape of the early Universe, but it was not the only game in

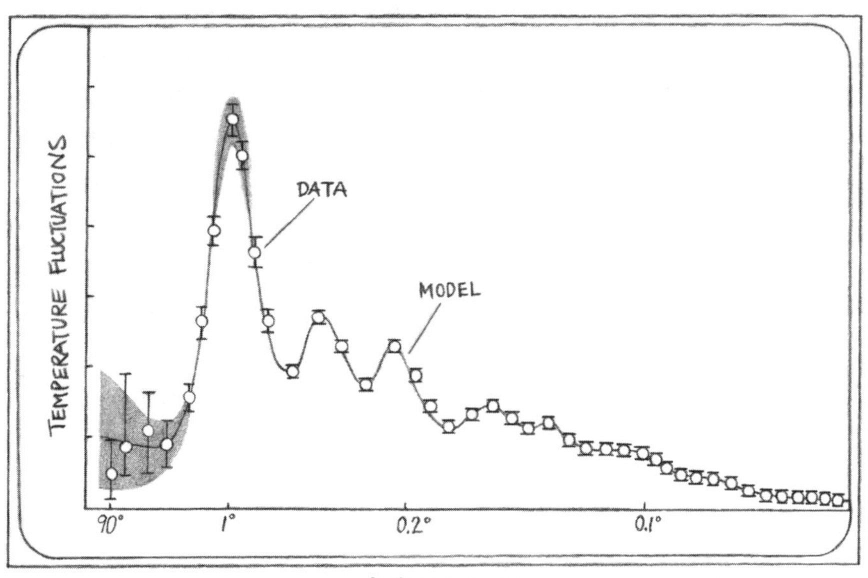

Cosmic microwave background temperature anisotropy power spectrum
The power spectrum encapsulates the amplitude of the tiny fluctuations of the CMB temperature on different angular scales on the sky. The data allow us to constrain parameters in our current cosmological model.

town. Consider one of those slightly higher-density regions in the primeval plasma. The presence of more mass in this environment means a stronger gravitational force acting on nearby matter, causing it to stream inwards, further increasing the local mass. Gravity causes this clump of matter to contract, increasing in density. You can think about it like a special type of well: one that gets deeper as it fills with water. The deeper the well, the stronger the gravitational pull on surrounding matter.

The important point is that the photons are coupled with the baryons, and so when baryons get dragged into the well through gravity, so do the photons. If gravity had its way, the matter in the well would continue to be compressed to ever-higher density, and more matter would fall in, causing a runaway collapse. But as the photon-baryon fluid is compressed, an outward 'restoring' force is exerted by the photons through a process called radiation pressure. The photons exert pressure on the electrons through their electromagnetic interaction: they push back on the normal matter as it is compressed through gravitational contraction. The effect is similar to the compression of a spring: the spring will exert a restoring force opposing the compression.

This radiation pressure opposes the gravitational contraction, resulting in the baryons and photons being pushed back out of the well. But as they get pushed back to lower density, the radiation pressure between the photons and baryons decreases, and once again they fall back in to the well through gravity. As they are compressed, the pressure goes up again, pushing the photons and baryons back out. The pressure drops and the cycle repeats.

You can see here an oscillation of the photon-baryon fluid as the two processes compete, with pressure waves in the photon-baryon fluid propagating outwards from the wells, analogous to

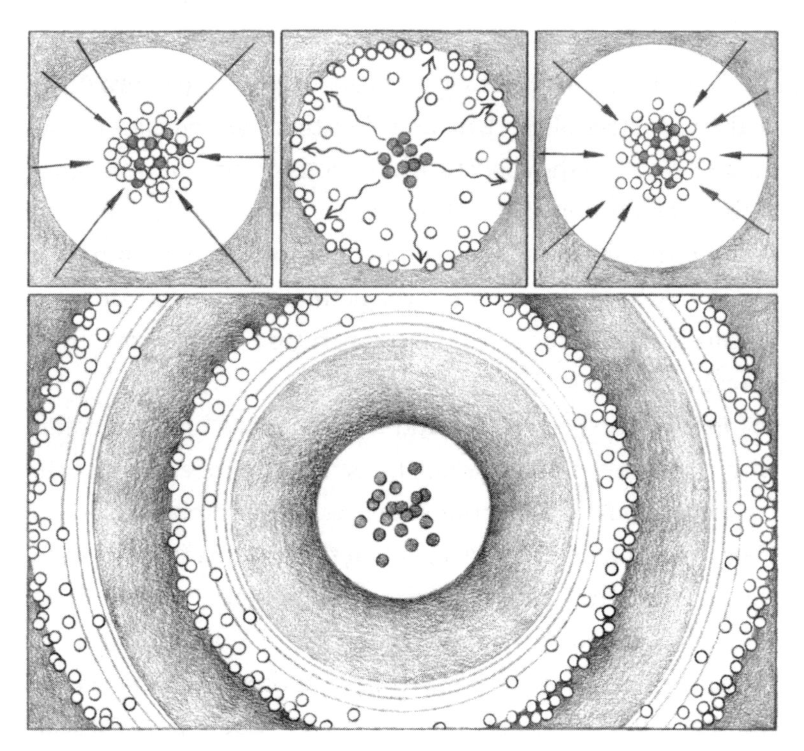

Baryonic acoustic oscillations

Baryonic acoustic oscillations in the photon-baryon fluid before the Epoch of Recombination. After recombination, the oscillations were frozen in place, and can still be detected in the distribution of matter today.

sound waves in air. It is no surprise then that these are known as 'baryonic acoustic oscillations'. They are spherical sound waves in the photon-baryon fluid: concentric shells of slightly enhanced matter density, emanating from their 'seed' points.

The acoustic oscillations can only occur up to the point of recombination. After that the coupling between radiation and matter is broken, and the photons stream away. With nothing to drive them, the expanding shells stall, and become frozen in place. It's a bit like throwing a pebble into a pond, allowing the concentric ripples to expand for a while and then suddenly turning the water to ice. So, not only are there fluctuations in the matter field seeded by primordial quantum perturbations at the very start of the Universe, but there are density fluctuations caused by the acoustic oscillations around these original seeds. The photons from this time that we now detect in the CMB keep the 'memory' of the temperature of the photon-baryon fluid at recombination, and so the pattern of the acoustic oscillations is also encoded in the temperature anisotropy power spectrum.

Just as the speed of a sound wave depends on the medium within which it is propagating, the speed of propagation of the baryonic acoustic oscillations was set by the properties of the photon-baryon fluid. It turns out that the speed of these acoustic waves was approximately half the speed of light. Fast, but finite. The combination of the finite sound speed and finite time the pressure waves could travel before the photons decoupled from the baryons sets an upper limit on the maximum physical scale of the acoustic oscillations. This is called the acoustic horizon, and has a diameter of about four hundred thousand light years. And that's where the big peak in the temperature anisotropy power spectrum comes from: it is known as the 'first acoustic peak' and corresponds to the largest scale that could be affected by the baryonic acoustic oscillations before the Epoch of Recombination. In the frozen pond, it's the size of the largest ripple.

We see other peaks in the temperature anisotropy power spectrum, and these represent the acoustic oscillations on smaller scales (the smaller ripples), but their amplitude is lower compared to the first peak, dwindling in power as one looks to finer scales. This 'damping' is due to the fact that recombination did not happen instantaneously: some of the photons still underwent scattering as electrons started binding to protons. In effect, the photons could diffuse out of small-scale high-density perturbations, and this smooths out the imprint of the acoustic oscillations in the power spectrum on the corresponding small angular scales.

Like all density fluctuations, the high-density shells left by the acoustic oscillations grew over time through gravity. Within these environments, galaxies had a better chance of forming. Our basic picture of galaxy formation is of baryons – mainly in the form of hydrogen atoms – being drawn into the gravitational 'wells' formed from clumps of dark (and normal baryonic) matter. When that gas reaches a sufficient density, it ignites into stars and a 'galaxy', as we know it, is born. Since the physical scale of the baryonic acoustic oscillations was frozen at the point of recombination, this led to a natural bias in the locations where galaxies formed – a higher probability within the denser shells. One might predict that it could be possible to see a signature of this in the large-scale distribution of galaxies even today.

In 2005 a project called the Sloan Digital Sky Survey, a large imaging and spectroscopic survey of stars and galaxies, measured the so-called 'two-point correlation function' of nearly fifty thousand galaxies in the relatively nearby Universe. The two-point correlation function is basically a measurement of the probability of finding pairs of galaxies at different separations compared to the number of pairs you would expect at the same separation if the galaxies were randomly distributed in space. The overall shape of the correlation function had been well established for many years, but the strength of Sloan was in the quality of data: it allowed for a very accurate measurement of the correlation function because of the large number of galaxies observed across a very large swathe of sky. The data revealed a characteristic 'blip' of excess probability – a propensity to find pairs of galaxies – at an angular scale that corresponds to the acoustic horizon at the time of recombination, but stretched to about 500 million light years due to cosmic expansion in the intervening billions of years. Remarkably, this is the imprint of the primordial baryonic acoustic oscillations, still present in the distribution of galaxies today.

That was a beautiful observational result, demonstrating that baryonic acoustic oscillations can be used for a deeper cosmological measurement: they can actually inform us about the expansion history of the Universe. If the oscillations persist throughout cosmic time, we can treat the fixed physical scale of the baryonic acoustic oscillations as a 'standard ruler' – a well-calibrated measuring stick. By measuring the apparent angular scale of the ruler at different redshift we can track the expansion history of the Universe, simply by comparing how much the ruler has been stretched by cosmic expansion. This is fundamental empirical data to help constrain our overall cosmological model, which makes firm predictions for the exact nature of the expansion.

Experiments are under way and still being planned to do just this. The key challenge is to map huge areas of sky to allow surveys to detect colossal numbers of galaxies; you need this to robustly detect and accurately measure the scale of the baryonic acoustic oscillations. The promise of using this technique to track the expansion history of the Universe has led to major international projects such as Euclid, a satellite designed to detect galaxies seen when the Universe was just a few billion years old. Euclid's primary mission – and, indeed, the goal of most experiments aiming to measure the evolution of cosmic expansion – is to help characterize the mysterious 'dark energy' thought to be responsible for *accelerating* the expansion of the Universe.

Evidence for the Universe's accelerated expansion first came from observations of distant supernovae, which appeared fainter than expected for a cosmological model with a constant or decreasing rate of expansion. A particular type of supernova, unimaginatively named 'Type Ia', are a class of 'standard candle': objects of known intrinsic luminosity. As with Cepheid variable stars, if you know the intrinsic luminosity of an object and measure its observed brightness you can determine its distance according to the inverse square law, the rule that dictates how faint an object should appear at a certain distance. You cannot see Cepheids in very distant galaxies – they are too faint – but you can see the supernovae at tremendous distance.

The problem is that stars, and supernovae in particular, are rather messy objects. Distance measurements based on Type Ia supernovae are subject to various complicated 'systematic' effects that must be carefully corrected or accounted for. To be blunt, there is much about the detailed astrophysics of supernovae that we do not yet understand, and this can introduce uncertainty in the distance measurement. Baryonic acoustic oscillations are far more elegant, in the sense that one simply makes a geometric measurement – the angular size of a feature on the sky – to track the expansion (hence the obvious connotation of the name of the Euclid mission).

I said earlier that the microwave background photons have enjoyed a nearly uninterrupted journey across the bulk of cosmic history, from the Surface of Last Scattering to our telescopes on and around Earth. This 'nearly' is important. You can think of the Surface of Last Scattering – the origin of the CMB – as a kind of backlight to the entire Universe. As we know, we do not live in an empty Universe: the space between us and the Surface of Last Scattering is filled with various structures that the photons encounter on their journey: galaxies, dark matter and intergalactic gas. As the photons travel through this landscape, subtle distortions are imprinted on the pure, primordial CMB blackbody spectrum. In other words, the energy distribution of CMB photons is slightly modified along the way. Rather than being a nuisance, we can actually use this to learn more about the astrophysics of those intervening environments.

The most famous example of such a distorting effect arises when a CMB photon travels through a cluster of galaxies. A cluster is a huge assembly of galaxies, perhaps numbering thousands of star systems, packed into a roughly spherical volume maybe just a few million light years across. These are the largest gravitationally bound objects in the Universe, formed at the highest peaks in the primordial density field. A cluster is permeated with a 10-million-degree plasma that engulfs all the galaxies within it. This hot atmosphere, called the intracluster medium, gets its thermal energy from the gravitational 'potential' energy of the system. The gas itself originates from intergalactic space (there is far more gas outside of galaxies than inside them), and this is attracted to the clusters through gravity. As the gas accelerates towards the cluster, it heats up. It gets so hot that the electrons get separated from the protons, and we are back to a medium in which we have energetic free electrons – a plasma.

This plasma emits its own radiation: X-rays emitted by the high-speed electrons as they pass by protons, a process called 'free–free' emission. This means we can detect clusters of galaxies not only as dense clumps of hundreds or thousands of galaxies, but through the telltale glow of X-rays emitted from the hot intracluster medium.

A CMB photon that happens to travel through a cluster has to contend with this plasma. And just as photons were scattered by free electrons before recombination, some CMB photons also scatter with free electrons in the intracluster medium. In this case a CMB photon can get a kick of energy, increasing its frequency. These kicks distort the near-pure blackbody spectrum of the CMB when there is a massive cluster in the way: the intracluster medium shifts some of the CMB photons to higher frequencies. The observational outcome is that, at certain observed frequencies, the CMB will appear to have a hole, or drop in flux, at the position of a cluster, and at other frequencies the CMB will appear brighter where there is a cluster. This is called the thermal Sunyaev–Zel'dovich effect, named after Rashid Sunyaev and Yakov Zel'dovich, the astrophysicists who first described it.

The wonderful thing about the Sunyaev–Zel'dovich effect is that the size of the effect scales with the total pressure of the electrons in the plasma filling the cluster, which in turn is related to the total mass of the cluster, dark matter and all. It means we can actually weigh these most colossal of cosmic objects simply by looking at their imprint on the cosmic microwave background.

There is another effect that, to some extent, affects all CMB photons during their journey through the Universe. Like the Sunyaev–Zel'dovich effect, we can use it to our advantage to learn something else about the contents of the Universe between Earth and the Surface of Last Scattering. It is exciting because, as a method of exploring the Universe, it is just becoming feasible to exploit, thanks to the high-quality maps of the CMB we can now produce. The effect is called gravitational lensing.

As we know, the Universe is full of matter, and all bodies with mass exert a gravitational force on other massive bodies. Isaac Newton first described gravity as a force that acts between two bodies, proportional to the product of their individual masses and inversely proportional to the square of the distance between them. This Newtonian description works just fine for many situations, but Albert Einstein later refined the picture and showed us how gravity can be described through a distortion of spacetime, the fundamental canvas of the Universe.

We are going to explore this idea more later, but the gist is this: spacetime is distorted by the presence of mass, and all photons travel on paths through spacetime. The trajectories of photons follow the distortions and therefore can be deflected as they pass by clumps of matter, like a galaxy. This is gravitational lensing and is analogous to the deflection of a ray of light passing through an optical lens.

Light from the Surface of Last Scattering shines through all the intervening matter in the Universe, so all CMB photons we detect have been gravitationally lensed to some extent. Like looking at a distant landscape through an imperfect window pane, we never actually have a direct, undistorted view of the background radiation. But if we can estimate the amount of gravitational lensing – how much the photons have been deflected – then we have a way of mapping the distribution of matter across the entire Universe.

Incredibly, this has been done. In recent years it has become possible to make maps of CMB lensing. You don't see the distortion just by looking at the temperature map – the effect of lensing is far too subtle to notice by eye. Instead, lensing imprints a subtle statistical signal that can be recovered through some clever processing of the data. For example, two CMB photons emitted from independent parts of the Surface of Last Scattering might be lensed by the same foreground structure, such as a cluster of galaxies. This 'correlation' leaves a signature that, with some ingenious filtering, can be extracted, allowing us to map the amount of lensing – and therefore the amount of intervening mass – across the sky. It is like mapping out the imperfections in the window pane by knowing something about the landscape beyond.

These are tremendously valuable maps, because gravitational lensing of photons is not just caused by the visible matter that we can actually see, but all the dark matter as well. Galaxies grew grew the invisible framework of dark matter, so linking CMB lensing maps with surveys of galaxies offers a new way to learn about how galaxies have formed and evolved within the – generally unseen – large-scale structure of the cosmos.

Just half a century ago we had our first glimpse of the afterglow of the cosmic fireball, providing us with compelling evidence that the Universe began with a bang. Today, CMB experiments continue to move forward in scale, sophistication and sensitivity. Breaking down the microwave background light into different components, particularly its different polarizations, is the current challenge of observational cosmology, and will allow us to investigate the nature of the Universe *immediately* after it formed. Polarization is a way of describing the coherence of the orientation of the oscillations of the electric and magnetic field of a set of electromagnetic waves in space. If the oscillations are in the same orientation for all the waves (or, if you like, photons) then we say that the radiation is polarized.

So-called 'B-mode' polarization describes a particular geometric configuration where the orientation of the oscillations of the electric and magnetic fields are at 45 degrees with respect to the direction of travel. Now, B-modes are not produced by the Thomson scattering that occurs between photons and free electrons near the Epoch of Recombination, but they *are* predicted to be introduced by gravitational waves – literal ripples in spacetime – during the brief period of cosmic 'inflation', or hyper-expansion, that is thought to have occurred right at the start of time.

Photons that became polarized in this way should be detectable today. They would provide much-sought-after observational evidence for a fleeting inflationary epoch at the start of the Universe, filling in an important piece of the cosmological puzzle. The catch is that the B-mode signal is expected to be extremely weak and must be carefully rescued from an almost overwhelming electromagnetic sea of contamination. To make matters worse, contaminating B-modes can be introduced by other sources, swamping the signal from primordial gravitational waves. But the challenge is worth the effort. This elusive signal, carried by light that has taken nearly the entire lifetime of the Universe to reach us, will reveal information about the mechanics of the very birth of the cosmos. We are on the verge of detecting it.

THREE

STARLIGHT

I grew up in Cornwall in the far southwest of England, in a village far away from the light pollution that can overwhelm the night sky close to towns and cities. Our back garden looked out across farmland, and just over the horizon was the sea. Nights could get wonderfully dark. I have fond memories of crisp, clear winter nights on the frosting grass, the sky awash with stars. What I loved most was the combination of the quiet stillness and almost unfathomable scale of the night sky, impossible to take in all at once. On the darkest nights the bright band of the Milky Way arched brightly overhead, hinting at some sort of structural order in the firmament. The harder I looked the more I saw.

When I think back to what led me into astrophysics as a career, starlight was undoubtedly my guide. Deep down I wanted to know what those pinpricks of light were – not just that they are stars, but what a star actually is. For me, that questioning has taken me on a journey of discovery as a scientist, a journey that has given me a more intimate understanding of nature than I ever thought possible. Yet I still feel I have only scratched the surface. The harder I look the more I see.

Observations of stars was our species' first insight into cosmology. Ancient civilizations arising independently across the planet began to chart the heavens: bright stars and the patterns they form on the sky – the constellations and asterisms – were given names and bestowed with mythology. How many of us are familiar with the signs of the zodiac: Aquarius, Taurus, Gemini, Sagittarius and so on? These are the constellations that happen to be crossed by the 'ecliptic', the apparent path of the Sun and planets across the sky, betraying the plane of the Solar System.

Most constellations are just chance groupings of stars at very different physical distances that happen to line up in projection, forming shapes that humans are particularly good at recognizing: typically combinations of lines, squares and triangles. But some groupings of stars that you can see are real physical associations, like the Pleiades. They are a cluster of bright blue stars also known as the Seven Sisters or, in Japan, Subaru, to be found just off the horn of Taurus. The stars in the cluster that you can easily see with your eyes are the brighter members of a clutch of stars that formed about one hundred million years ago. They are relative newcomers to the Milky Way; for comparison, our Solar System is fifty times older.

Without the aid of a telescope, for most of us stars are the only observable link to the Universe beyond our Solar System. Even so, on the darkest of nights we are only able to see between a few and ten thousand stars with the naked eye, and most of these are in our Solar neighbourhood. This is the tip not just of an iceberg but of an ocean of icebergs. For while there are a few hundred billion stars in our own galaxy, there are at least several hundred billion – perhaps as many as a few trillion – other galaxies in the observable Universe.

We know about the existence of other galaxies now, but for most of human history the stars above were thought to be the Universe entire. Some models had the stars embedded on the surface of a 'celestial sphere', or nests of spheres, around us. Other models had the stars spread throughout all space around the Earth. Although some natural philosophers and scientists had proposed it even as early as the seventeenth century, it was only in the first quarter of the twentieth century that it was empirically determined that the Milky Way is its own star system, separated by great distance from its nearest counterpart: the 'Great Galaxy' in the constellation of Andromeda, usually just referred to as the Andromeda Galaxy. I said that stars were generally the only naked eye link to the Universe beyond our Solar System, but that's not quite true. If you find yourself under a very dark sky and have keen eyes, you can just about make out the faint smudge of light that is the Andromeda Galaxy, hiding among the foreground stars of the Milky Way.

It was the determination of the distances to Cepheid variable stars in the direction of the 'spiral nebula' (as it was then known) in Andromeda by Edwin Hubble in the early 1920s that showed that they lie much further than the edge of the Milky Way. When Henrietta Swan Leavitt's period-luminosity law for Cepheids was applied, which allowed Hubble to estimate their distance, they were found to be about forty times further away than the most distant stars in our galaxy. This observation showed that the Cepheid variable stars in the spiral nebula in Andromeda reside in a separate, far-flung star system.

This can be considered the birth of 'extragalactic' astronomy and marked a sea change in the way humans pictured ourselves in the cosmos. Look in any direction you choose, save for the dense plane of the Milky Way itself (which is so thick with stars, gas and dust that it is impossible to see through), and you will find a Universe that is teeming with galaxies. And perhaps the most obvious characteristic of all galaxies is their stars.

So we should ask a simple question: what *is* starlight? Well, the first thing to say might sound obvious, but it is worth saying anyway. We humans can *see* starlight. That means stars emit radiation in the visible part of the electromagnetic spectrum. Now, that's not a coincidence of course. Life on Earth evolved as it was being bathed in the glow of our Sun, a source of energy that allowed most biology to develop. Plants evolved a method to turn carbon dioxide and water into glucose and oxygen using sunlight in a process called photosynthesis. It was photosynthesis that provided much of the atmospheric oxygen that animal life on Earth evolved to respire. Many species, including our own, have evolved receptors – eyes – that can detect electromagnetic radiation across a range of energies that closely match the peak range of energies of photons emitted by the Sun. Obviously this has fantastic advantages for survival: being able to spot a predator because some of the Sun's light reflects off its body and into your eyes, to be decoded into an image in your brain, is a rather useful skill. So most life as we know it on Earth is intimately linked to the properties of sunlight. We should expect life on other planets to be tailored to suit the radiative properties of whatever type of star they happen to orbit.

What else can we note? Well, when you go outside on a sunny day, you can see colours everywhere: the blue sky and the yellow-white sand, luscious green grass, fluorescent graffiti, pink-purple sunrises, iridescent dragonflies and grubby black dustbins. All those colours and their myriad shades are contained within sunlight. They arise because the Sun emits radiation across a broad spectrum of energies. As we have learned, the energy of electromagnetic radiation is characterized by the frequency of the wave, or the photon, carrying that energy. When waves of different energies interact with different media, they are transmitted (the physics term for passing right through), absorbed, reflected or scattered in different ways depending on the properties of the medium and the frequency of the light. It is the countless interactions between light and matter on the quantum scale – between the photons and the atoms and molecules that make up materials – that give rise to the fantastic variety of texture and colour we observe around us.

Let us consider the path of light from the Sun to the Earth. The first thing sunlight encounters when it arrives at Earth after its eight-minute, one-hundred-million-mile journey across the inner Solar System is our atmosphere, which is mostly a mixture of molecules of nitrogen and oxygen. All daylight originates from the same place – the Sun – and we can see the solar disc clearly as a well-defined entity in the sky. So why do we see light all around us during the day? Turn your back to the Sun and the sky in front of you is still bright blue. Why is this?

On a clear day we see the sky as uniformly blue because the molecules in the atmosphere are scattering the solar photons. We have already encountered light scattering: it means that the photons interact with particles in such a way that they get randomly redirected on to different paths. For sunlight travelling through the atmosphere, the scattering is more pronounced for photons of a certain energy: as you increase the energy of the photon (make its colour bluer), its chance of being scattered by an air molecule increases. Rather than travelling through the atmosphere on a direct path, this scattering causes sunlight to get bounced around all through the atmosphere, with a single photon undergoing many scattering events before eventually 'escaping' to our line of sight into our eyes. It is the blue photons that get scattered the most through the atmosphere, and this makes the sky appear blue in all directions. This process is known as Rayleigh scattering, after the physicist Lord Rayleigh (1842–1919).

Interesting things happen when the Sun rises and sets. You will have of course noticed that at dawn and dusk the sky isn't as blue as it is at midday; the horizon can blaze spectacularly with oranges and pinks and reds. These are colours towards the low-energy end of the visible light spectrum. If you think of the atmosphere as being a thin shell of gas around the solid globe of the Earth, when we look in different directions from the surface we are seeing through different path lengths of atmosphere, or in astronomy-speak, through different 'airmass'. If you look straight up to the zenith then you are looking through the lowest possible airmass, and when your view grazes the horizon it is the highest airmass – there is simply a longer column of air between your eye and the edge of the atmosphere. The more atmosphere that sunlight has to pass through to reach us, the higher the chance that the photons are scattered. In the extreme, the bluer, higher-energy photons are scattered so much that they never actually reach us. Only the longer wavelength photons – the redder ones – get through. That's why sunsets are red.

The fact that the 'sky' looks black in space is because there is no scattering medium. The Sun is still shining of course, bathing objects such as spacecraft and the surface of the Moon with light, but with no diffuse medium of atoms or molecules to interact with, any photon that does not hit a piece of matter – such as an astronaut's visor – is simply lost to the Universe, streaming away into space, maybe one day to be seen by observers on a distant planet; to them just another faint star in their sky.

Indeed, the fact we can detect *other* stars with our eyes means that they too are emitting electromagnetic radiation in the visible part of the spectrum. But if you look closely, you may notice that the stars are not all the same colour. At first glance they all appear bright white, but if you let your eyes become accustomed to the darkness and flick your gaze between some of the brighter stars, you will notice a difference in colour: some appear quite red, others blue. A long-exposure photograph will really bring out the difference, but the colours of the stars are definitely possible to discern with the naked eye. There's a great quote about this that I love from George Orwell's *Down and Out in Paris and London*. Our protagonist meets the street artist Bozo, who exclaims, 'Say, will you look at Aldebaran! Look at the colour. Like a – great blood orange!' It is surprising if you have never noticed that the stars have colour, but the colours of stars aren't just pretty, they hold important clues about their astrophysics.

Why are stars different colours? The short answer is quite simple: temperature. Think about heating a metal rod with a blow torch. As the rod heats up it starts to glow red, then orange, then blue. It is emitting electromagnetic radiation in the form of visible light. Within the metal rod are countless charged particles, the electrons, binding together the more massive atomic nuclei, which are all arranged in a lattice-like structure. The fire of the blow torch increases the thermal energy of all of these particles, causing them to vibrate and oscillate.

As we know, for a charged particle surrounded by an electrostatic field, any oscillation of the particle will also oscillate the field. The unification of electricity and magnetism tells us that the oscillating electric field generates an associated oscillating magnetic field, which produces an oscillating electric field, and so on. These oscillations propagate away from the particle as an electromagnetic wave. As the energy increases, the particles oscillate faster and so the frequency of the radiation also increases. The bar glows from dull red to blue-white, until the thermal energy of the particles – the temperature of the bar – becomes high enough to break the bonds holding the particles in the metal together, and it melts.

We can relate this physics to the stars: blue stars are hotter than red stars. Specifically, we are talking about the surface temperature, that of the photosphere – the outer layer of the star – where photons are actually released into space. The spectrum of light emitted by a star is broadly characterized by the same 'blackbody' shape we encountered in the cosmic microwave background. Unlike the cosmic microwave background, stars are far from perfect blackbodies. In part this is because they are rather complex and messy objects, but nevertheless the blackbody spectrum is a decent 'first order' approximation to the light they emit. The stellar spectrum spans a broad range of energy, but with a characteristic peak in 'intensity' at a specific frequency, and that frequency is determined by the star's temperature. This determines the colour of the star.

Stars come in a range of temperatures, and therefore different colours, because not all stars are born equal. To be more specific, stars are not all formed with the same mass, and it is stellar mass that is the key determinant of a star's properties and evolution.

Take a large group of randomly chosen people and measure their weights. When you count up the number of people in different intervals, or 'bins', of weight and plot the distribution, you will tend to find a characteristic bell-shaped curve, called the 'normal' distribution. It peaks at some average weight but there is a symmetric spread around this, tailing away at the extremes. Most people are close to the average weight of the sample, and there are very few extremely under- or overweight people.

Take a group of stars recently formed from the same cloud of gas and measure their masses, and you will also find a characteristic distribution. It's not a bell-shaped normal distribution in this case: we tend to find lots of low-mass stars down to some critical mass threshold (below which there are no stars), and few very massive (say, above ten times the mass of the Sun) stars. The shape of this distribution is called the stellar initial mass function. It describes the number of stars born with different masses from a cloud of gas. Determining the shape of the initial mass function, particularly at its extreme ends, remains a hot topic of research. In part this is because the astrophysics that determines the shape of the initial mass function – the very formation of stars themselves – is not fully understood, and yet it is fundamental to our understanding of the astrophysics of galaxies.

We know there is a strong relationship between the mass of a star and the rate at which it radiates energy, its luminosity. Simply put, the most massive stars are the most luminous, and low-mass stars are the least luminous. Related to this, there is also a link between the luminosity of the star, its surface temperature and its radius. We already know that the temperature of a star affects the peak frequency of its blackbody-like spectrum. As the temperature increases, the blackbody peak skews towards higher frequency. When this happens, the total amount of energy radiated per unit surface area of the emitting body also increases. So hotter blackbodies radiate more energy than cooler ones.

The relationship between the luminosity of any blackbody and its physical size and its temperature is known as the Stefan–Boltzmann Law. This states that the total luminosity of a blackbody is proportional to the fourth power of its temperature. Basically this means that a small increase in temperature will result in a big increase in luminosity. Double the temperature and the luminosity increases by a factor of sixteen. The radius of a star is also largely determined by its mass: massive stars are bigger. When we put all these things together we find that, generally, the most luminous stars are massive, physically large and blue, and the faintest stars are low mass, small and red. This means that galaxies are filled with stars of different sizes, luminosities and colours.

NEUTRON STAR ~10 KM

WHITE DWARF 0.01 R⊙

PROXIMA CENTAURI 0.15 R⊙

SUN 1R⊙

BELLATRIX 6R⊙

BETELGEUSE ~1000 R⊙

But where does starlight actually come from? How is it produced? Stars might be approximated as simple spherical bodies of different sizes, emitting radiation from their surfaces, but in reality all stars are balls of hot gas with complex astrophysical processes going on deep inside. The surface of the Sun, the photosphere, is a seething, violent plasma where colossal, planet-dwarfing prominences and flares erupt from the 6,000-degree surface, drawn out by churning magnetic fields. As with all stars, light escapes (or, if you like, 'shines') from this surface, but the main action happens down in the stellar core. In the core the pressure and temperature of the gas are much, much higher than the surface. It is where nuclear fusion – the power source of stars – occurs.

Comparison of stars and stellar remnants

Betelgeuse (distance 725 light years) is a red supergiant; is has exhausted the hydrogen in its core, expanded in size and cooled; it is approaching the end of its short life, soon to explode as a supernova.

Bellatrix (distance 250 light years) is about nine times the mass of the Sun; it is hotter and more luminous than the Sun, appearing as a bright blue-white star in the constellation of Orion, fairly close to Betelgeuse on the sky.

The Sun is a fairly typical main sequence star, about half-way through its 10-billion-year lifespan.

Proxima Centauri (distance 4 light years) is a red dwarf star, just over a tenth of a solar mass. It is a low-mass main sequence star, with a surface temperature about half that of the Sun, giving Proxima a red hue.

White dwarfs and neutron stars are the compact remnants left behind after star death. White dwarfs are most common; they have a mass similar to the Sun, but a size similar to the Earth. Neutron stars are left behind after the explosive deaths of the most massive stars, and are even denser than white dwarfs. Both white dwarfs and neutron stars are held up against gravitational collapse by the 'degeneracy pressure' that opposes electrons and neutrons from being compressed together; a result of the Pauli Exclusion Principle.

Like all stars, the Sun formed from a cloud of gas, mainly made of hydrogen. Hydrogen is the lightest and most abundant element in the Universe, composed simply of one proton and one electron. Virtually all atoms of hydrogen in the Universe formed when electrons combined with protons nearly four hundred thousand years after the Big Bang at the Epoch of Recombination. Dark matter aside, hydrogen is the raw material of galaxies and all the normal 'baryonic' matter within them.

For stars to form within a cloud of hydrogen, parts of the cloud must contract to form high-density clumps. This can occur – provided there is enough mass present in the same spot – through gravitational collapse. In any gas, be it astrophysical or terrestrial, the particles are not bound together with electrostatic bonds. Instead, they are whizzing by each other at high speed. There is a lot of internal kinetic energy in this bundle of particles. So the particles in the gas must first lose their kinetic energy to allow them to get close together. For a cloud of hydrogen, this can happen through collisions between the atoms. A collision can give a kick of energy to electrons within the atoms, which then release that energy in the form of a photon. If it is not intercepted and absorbed by another atom, the photon can escape through the gas cloud, carrying the energy away. The cloud gets a little cooler – the particles are not moving so fast – as a result.

We describe this as 'gravitational cooling'. Somewhat paradoxically, gas must cool before a star can ignite. If a gas of atoms of hydrogen can cool sufficiently, pairs of atoms can form bonds between themselves to form molecules of hydrogen: two atoms bound together. The bonding reaction can work like this: a single hydrogen atom can combine with a free electron to form a negatively charged hydrogen ion. This ion goes on to bind with another neutral hydrogen atom, detaching an electron in the process. This forms a single molecule of two hydrogen atoms bound together by the two electrons – one per atom – that are shared by the two protons in a covalent bond.

Another way that molecular hydrogen can form is on the surface of grains of interstellar dust, where more complex, but more efficient, chemical reactions can bind atoms of hydrogen together into molecules. In galaxies like the Milky Way, most of the molecular hydrogen tends to collect in huge structures called

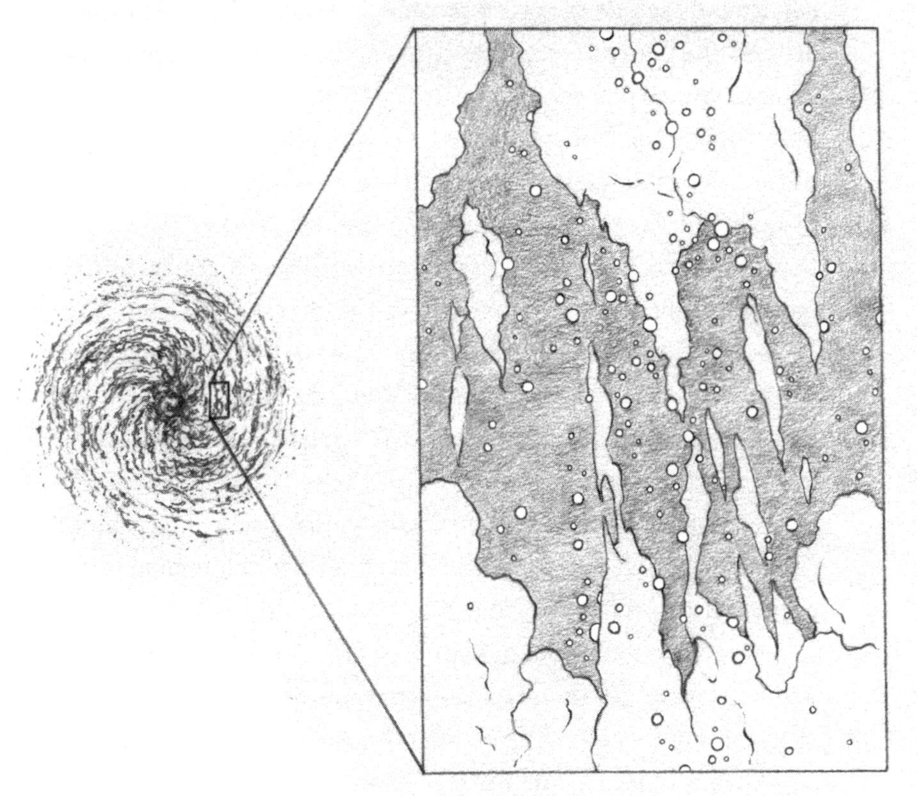

Star-forming region

Stars generally form within giant clouds of hydrogen gas. In a galaxy like the Milky Way, the majority of those clouds litter the galactic disc, and are mainly distributed along the spiral arms.

giant molecular clouds, peppered along the spiral arms of the disc of the galaxy. Each cloud is a colossal reservoir of gas, spanning up to hundreds of light years across and containing the equivalent mass of millions of Suns in molecular hydrogen. It is within these clouds that most stars are born.

What impels the formation of a star? If we could map out a giant molecular cloud in detail we would find that the distribution of gas is not perfectly smooth but filamentary and clumpy. The clouds are not static, either: they are moving with the general rotation of the galaxy, getting twisted and sheared in their orbits around the central hub. They are also buffeted by supernova explosions and powerful winds driven from the surfaces of stars, both outside and within the cloud. All this can introduce cascades of turbulence propagating through the cloud, creating filigrees of slightly denser gas, like a hand waved through a pall of smoke. Some of these dense clumps are susceptible to gravitational collapse if the local pressure due to the thermal energy of the gas is insufficient to balance the gravitational force due to the clump's mass. That threshold is called the 'Jeans mass', after physicist James Jeans (1877–1946). Crossing this threshold is the first stage of star formation.

As clumps of gas detach from the wider cloud and gravitationally collapse, they increase in density, and the temperature once again starts to rise. But in order to ignite nuclear fusion, two protons – the nuclei of hydrogen atoms – must be brought within intimate proximity for the attractive 'strong' nuclear force to fuse them together. The fusion releases energy in the form of a photon. But the strong force only operates over a very short distance, of the order of just 1 femtometre, or a millionth of a billionth of a metre. To help visualize this: if the distance between the Earth and the Sun was scaled down to 1 metre, 1 femtometre would still only be just over a tenth of a millimetre. The catch is that in order to achieve this close separation, the repulsive force between the two positively charged protons must be overcome. This is called the Coulomb barrier, and is a bit like an imaginary wall dividing two protons.

We can try to work out what sort of energy this requires by comparing the thermal energy of the particles, which is proportional to their temperature, to the electrostatic 'potential' energy at the scale of the proton radius, which is roughly the distance where the strong force kicks in. It turns out that the temperature you need to overcome the Coulomb barrier – to climb over the wall – works out to be nearly ten *billion* degrees. Now, this temperature is not actually achievable in realistic conditions for the gas: the particles simply cannot reach that sort of thermal energy. Nevertheless, stars shine. So how does fusion actually get started?

The answer lies in quantum mechanics. As we know, quantum theory established that particles such as electrons and protons can behave like waves, and in turn, electromagnetic waves can behave as particles, the photons: wave–particle duality. We can think of a single particle, like a proton, not as a distinct point-like entity but as a kind of fuzzy cloud. The shape and density of the cloud is described by the wave function and represents the probability of the proton existing at a particular point in space and time, or more technically, in a particular quantum state. Basically, until it is measured, a particle is not definitely in one place, but *maybe* in many places.

This probabilistic nature of reality at the quantum level allows for things to happen that would not be possible if classical 'billiard ball' physics held at all scales. When two protons get close together, part of their wave functions can overlap, breaching the Coulomb barrier holding them apart. The barrier does still have an effect: if a proton is approaching the Coulomb barrier, its wave function will have a lower amplitude on the other side. In other words, there is a lower probability that the proton is on the other side of the wall. Low, but not zero, and that's the key.

This quantum effect is called 'tunnelling', because it is as if the proton has tunnelled through the wall. If there is a small probability that two protons can get closer together than the Coulomb barrier classically allows, and we have a huge number of protons, then in some cases the strong force can take hold, fusing together the protons and releasing energy in the process. Think of it this way: if I buy a lottery ticket there is a very small chance I am going to win the jackpot, but there is usually a winner every week. It is this quantum effect that kick-starts the

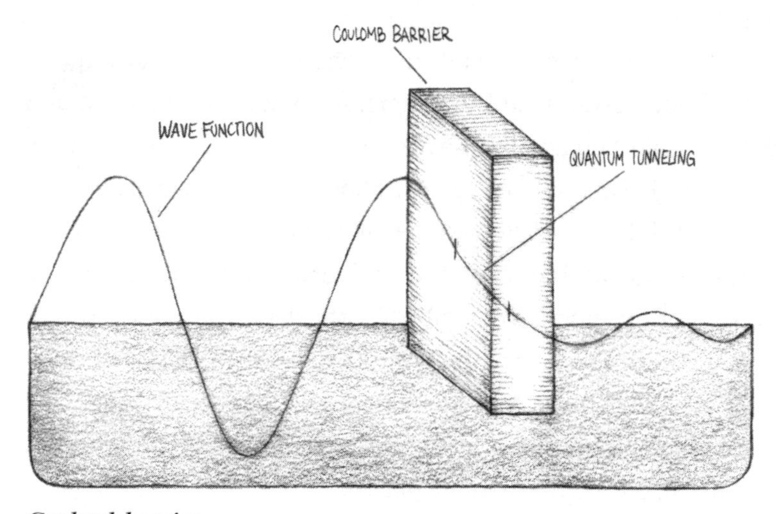

Coulomb barrier
Quantum tunnelling is an effect that allows the proton wave function to pass through the 'Coulomb barrier' that classically opposes two protons from getting too close together. This enables nuclear fusion to occur at a much lower temperature than is predicted by classical physics.

life of a star, allowing nuclear fusion to happen at a much lower temperature – albeit still millions of degrees – than predicted in the classical, non-quantum limit.

In a star like the Sun, a cycle of nuclear reactions takes place in the core that 'burns' hydrogen, forming helium as a by-product. That process is called the proton–proton – or 'pp' – chain. First, two protons react to form deuterium, which is a proton and neutron bound together, and this releases a particle called a positron (the antimatter opposite number of the electron) and a particle called a neutrino. Another proton can then bind with the deuterium to form an 'isotope' of helium called helium-3, which consists of two protons and one neutron (an isotope is a variation of an atom where the number of neutrons differs from the norm). This step releases energy in the form of a photon. Two particles of helium-3 can then combine to form helium-4, which is the regular, stable form of helium, releasing two protons. And so the cycle continues.

For stars more massive and hotter than the Sun, a slightly different nuclear reaction dominates the hydrogen burning, this time utilizing catalysts. This is called the carbon–nitrogen–oxygen cycle. This time the reaction goes like this: a free proton reacts with a carbon nucleus to form an isotope of nitrogen called nitrogen-13, releasing a photon. The nitrogen-13 is unstable, meaning it decays, forming an isotope of carbon called carbon-13 and releasing a positron and neutrino. Then, a proton reacts with the carbon-13 to form a regular, stable nitrogen nucleus, which also releases a photon. Next, another proton reacts with the nitrogen to form an isotope of oxygen called oxygen-15, and again releases a photon. The oxygen-15 decays to nitrogen-15 (yet another nitrogen isotope), emitting another positron and a neutrino. Finally, another proton reacts with the

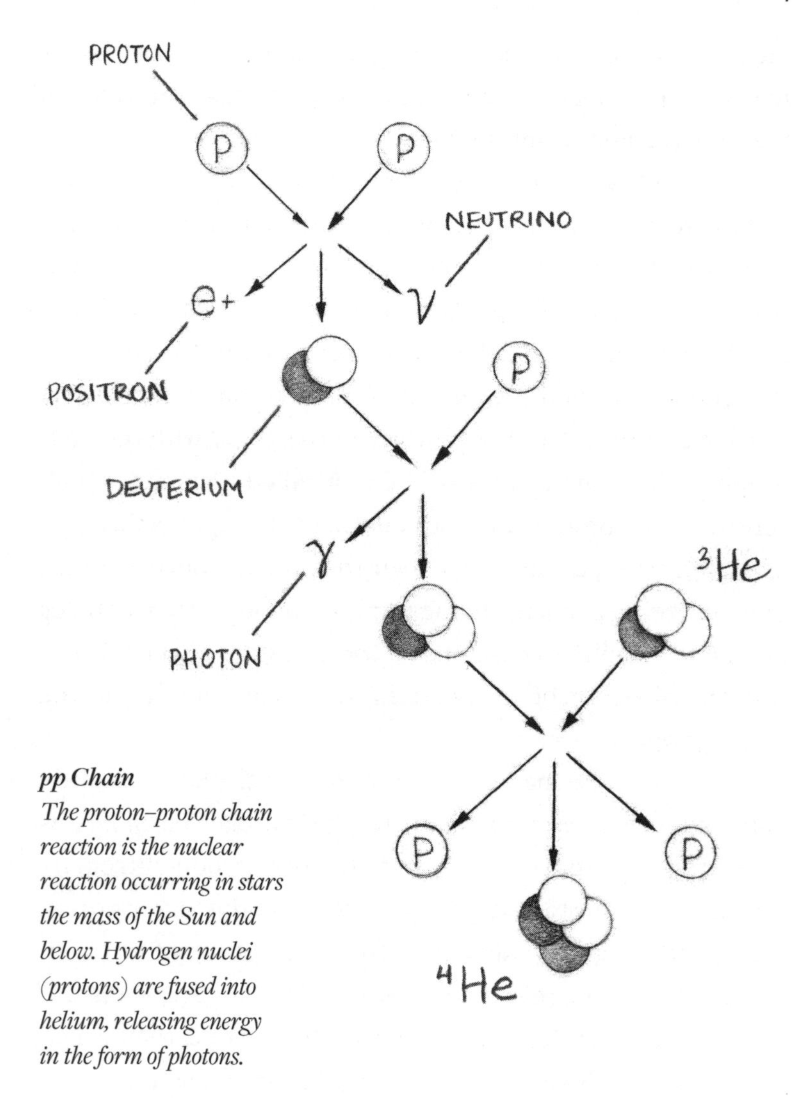

pp Chain
The proton–proton chain reaction is the nuclear reaction occurring in stars the mass of the Sun and below. Hydrogen nuclei (protons) are fused into helium, releasing energy in the form of photons.

nitrogen-15 to form regular carbon, plus a helium nucleus, and the hydrogen-to-helium burn is complete. Phew!

Hydrogen burning in the cores of stars produces photons – energy – as a product of the nuclear reactions. As each photon is produced, it simply tries to race away, but inevitably the photons interact with the surrounding gas, and this exerts an outward

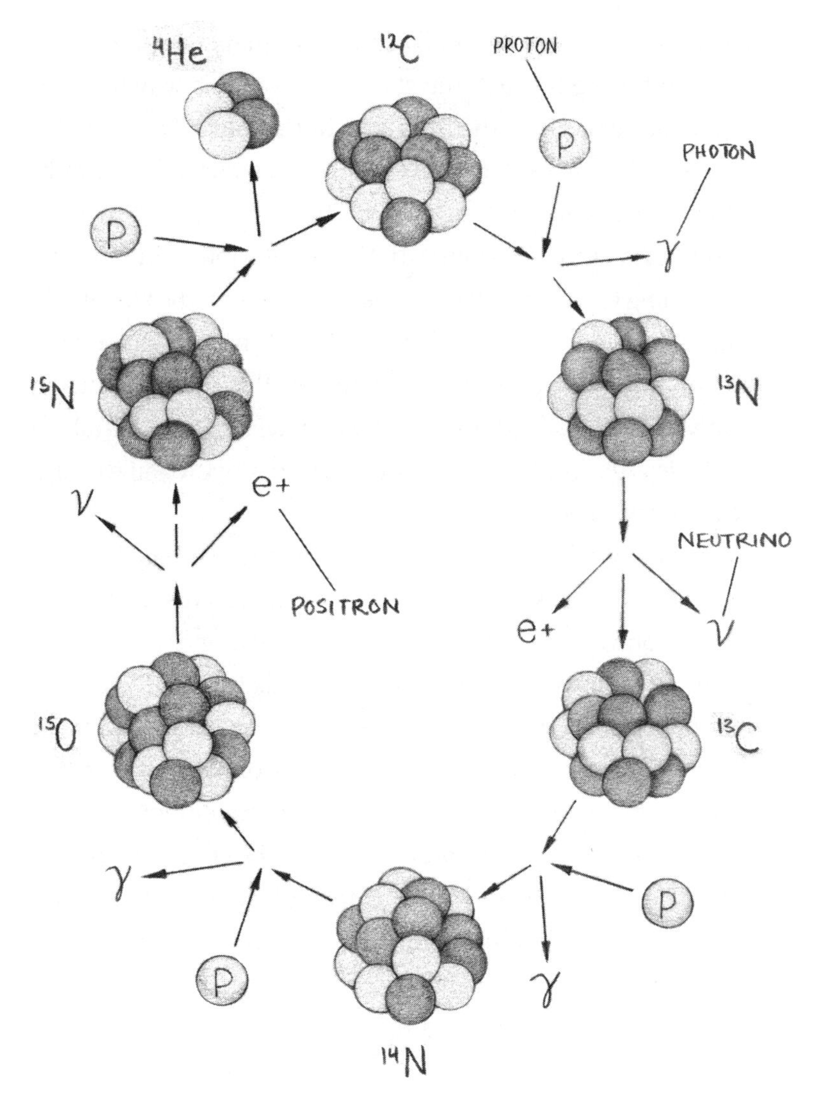

CNO Cycle

The carbon–nitrogen–oxygen cycle is a sequence of nuclear reactions that occur in the cores of stars more massive than the Sun. These elements catalyse the burning of hydrogen to helium and the process releases energy in the form of photons.

radiation pressure. At the same time, the energy from nuclear fusion is diffusing through the star itself, which is why the bulk of the stellar mass is in the form of a hot plasma. The particles within this plasma exert a thermal pressure, which, like the radiation pressure, also acts outwardly. The radiation and thermal pressure form an equilibrium with the gravitational contraction. With nothing to oppose gravitational contraction, the star would quickly collapse, and so the radiation and thermal pressure effectively hold the star up from the inside out. But this equilibrium can only be sustained as long as there are nuclear reactions taking place. This is generally only as long as there is hydrogen to burn; the fuel in the tank.

The rate at which a typical star consumes its gas depends on its mass: the most massive stars can sustain fusion for just a few million years, whereas the least massive ones can prevail for many billions of years. In any case, while hydrogen burning is taking place we say that a star is on the 'main sequence' of stellar evolution, referring to the tight relationship between the luminosity of a hydrogen-burning star and its temperature.

Once hydrogen burning is exhausted in the core of a Sun-like star, nuclear reactions will start to take place in shells around the core. During this process, the star starts to move off the main sequence, physically expanding in size and cooling to become a 'red giant'. With hydrogen burning in the core extinguished and the supporting pressure gone, the core collapses under gravity and the central density and temperature suddenly skyrockets. Eventually, the thermodynamic conditions allow the helium that has accumulated in the core of the star during its main sequence lifetime to ignite. We call it the 'helium flash'.

The helium flash reaction involves the fusion of two helium-4 nuclei into an element called beryllium, releasing a photon in the process. The beryllium nucleus then reacts with another helium-4 nucleus to form stable carbon-12 and another photon. Helium nuclei – two protons and two neutrons bound together – are also known as 'alpha' particles, and so the name given to the nuclear reaction that allows dying stars to burn their helium ash is the 'triple alpha process'.

Some of the carbon produced in the triple alpha process goes on to react with helium to form oxygen nuclei. And when that core is exhausted of its helium, helium burning also proceeds in a series of shells around the core. Towards the end of the process, the outer layers of the star begin to be ejected out into the interstellar medium, forming a nebula of glowing, expanding diffuse gases. We see these peppered throughout the disc of our galaxy. Left behind at the centre of the nebula is a dense, compact sphere of carbon and oxygen called a white dwarf.

This book is all about light, but it would be remiss not to mention something about all those neutrinos that are produced alongside the photons in stellar fusion. Neutrinos are extremely light particles that rarely interact with other matter and travel nearly at the speed of light. And they simply pass straight through anything in their way. Unlike the photons, neutrinos are generally not beholden to any interaction with matter. They just stream out from the core, through the innards of the star and away into the Universe. It follows that this steady flow of stellar neutrinos is also passing through you and me all the time, day and night – it doesn't even matter if the Earth is in the way. Tens of billions of them pass through every square centimetre every second. It means that there are over one hundred thousand solar neutrinos in your body right at this moment. Like phantoms they pass through unnoticed, with you for a moment before disappearing away into the Universe.

The photons being produced in nuclear fusion reactions in the core *do* interact with the surrounding stellar matter in a way we have already encountered: scattering with free electrons. For a star to actually shine, energy – the photons – must diffuse from the core of the star to its surface. One of the main ways for this to happen in the interior of stars like the Sun is through a process called radiative diffusion.

When we talked about the mean free path of photons at the time of recombination in the early Universe, we were talking about the average distance a photon travelled before interacting with an electron and scattering off in a random direction. In the centre of a star the particle density is so high that the mean free path of a photon is tiny – just a centimetre or so. This means that any fresh photon produced in a fusion reaction cannot simply go directly from the core to the surface and escape into space as starlight. Instead, it must undergo a long 'random walk', continuously scattering around inside the star, eventually to find the surface. This is no short stroll: it takes a photon anywhere from a thousand to ten million years to get from the core to the photosphere. It is a bit like being lost in the centre of a forest and trying to find your way to the edge by walking in straight lines, randomly changing direction every time you meet a tree.

This process of diffusion allows energy to be transferred from the centre of the star to its outer layers. But only a small fraction of the photons inside the star actually escape as starlight – most are continually absorbed, re-emitted and absorbed again in the huge reservoir of hot plasma below the photosphere. It is these interactions between the photons and the charged particles in the plasma that determines the broad distribution of photon energies that eventually do escape the star. The thermodynamic properties of the plasma – the distribution of particle speeds inside the star, in effect – are imprinted on the emergent photon energies as the blackbody spectrum, the same way that the range of energies of the cosmic microwave background photons are imprinted on them by the thermodynamic properties of the plasma before their last scattering. In both cases the matter and radiation are in thermal equilibrium before the photons escape.

If stars are great balls of fire getting ever hotter and denser toward their cores, why do we see the Sun as a distinct sphere with, at first glance at least, a clearly defined and solid-looking surface? Again, this comes back to the concept of the mean free path of photons and the optical depth of the medium they are passing through. The photons generated deep in the core have to diffuse out of the Sun over thousands or even millions of years, undergoing countless scattering events. As a photon nears the surface of the star, the gas becomes more diffuse and the mean free path – the average distance travelled between interactions with electrons – increases. At some particular distance from the core, the photon will undergo one last scattering event before the mean free path becomes, effectively, infinite and the photon escapes in a straight line. Some of those straight lines intercept Earth.

For a given star, the distance from the core where this last scattering occurs is the same in every direction, so we see the Sun as a round ball. What we observe as the surface of the star is where the photons are escaping from their own 'surface of last scattering', beyond which the optical depth is too high for us to see through. You'll recognize that this is exactly the same principle as the Surface of Last Scattering that appears to emit the cosmic microwave background.

Despite the long time it takes for any one photon to escape a star, the Sun still shines brightly in our sky. It lights our day

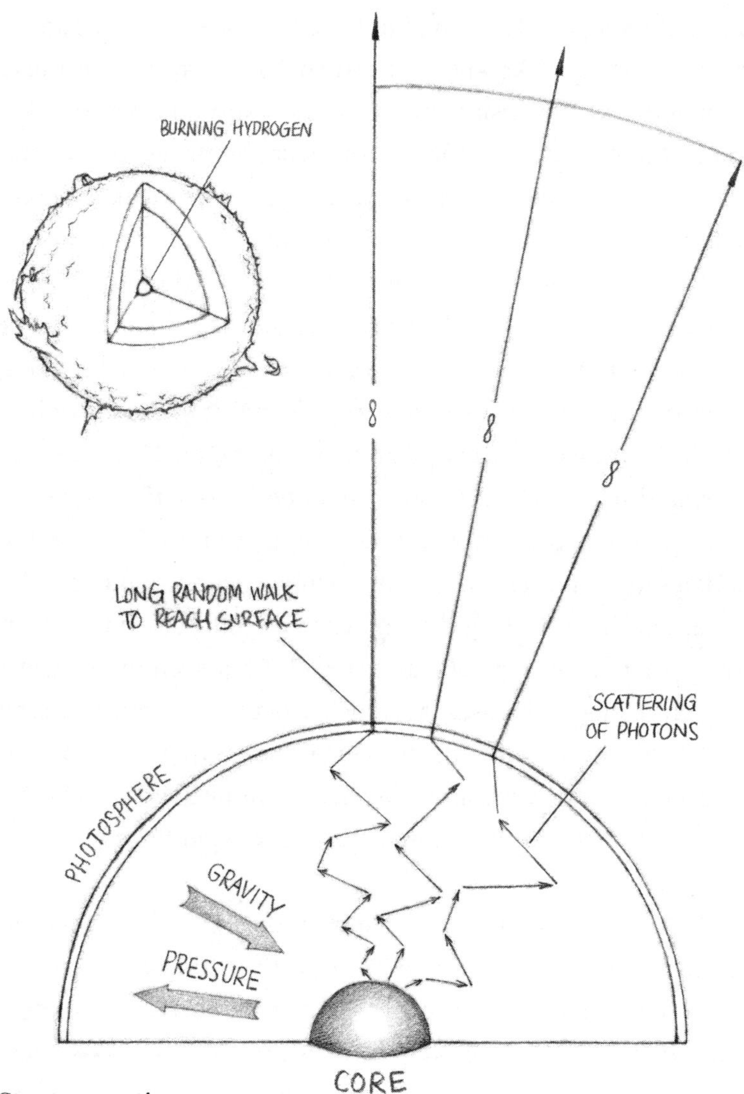

Star cross-section

Stars are balls of hot gas. In their cores nuclear fusion 'burns' hydrogen into helium, releasing electromagnetic energy that diffuses out through the star. Thermal and radiation pressure exert an outward force on the gas in balance with the gravitational force, which is trying to collapse the star. Photons generated during fusion can take millions of years to diffuse from the core to the surface, undergoing countless scattering events before finally escaping into the Universe as starlight.

and we can feel its warmth. But this is only because the Sun is very large compared to the Earth and we are, astrophysically speaking, right next to it. The amount of solar energy hitting the Earth during the day is about 1,400 watts per square metre (the exact value changing depending on how far you are north or south of the equator). This is called the solar irradiance. That stable flow of energy is sufficient to keep the Earth adequately warm and flooded with light to sustain an ecosystem that has, in one form or another, prevailed over four billion years or so. But it does not take a long trip into the Universe for sunlight to fade into insignificance.

Out at Pluto, which is forty times further away from the Sun than the Earth is, daylight pales. At Pluto, solar irradiance falls to less than 1 watt per square metre. This is the unforgiving inverse square law at work: the 'flux' of a luminous object – that is, the amount of energy flowing through a unit area (say, a square metre) per unit time (say, every second) – declines with the square of the distance from the radiating object. If the Sun was as far away as the next nearest star, Proxima Centauri, then the solar irradiance at the Earth's surface would be about twenty billionths of a watt per square metre.

You can see that the amount of energy falling on Earth from distant stars in our own galaxy is tiny. And from all the stars in all the other galaxies, it is fantastically small – almost negligible. And yet we *can* see the stars, and we can build telescopes to collect and record their light, even if that light has travelled for most of the age of the Universe to reach us.

Regardless of its source – within our Solar System or galaxy or far beyond – after travelling across space and time, all starlight we see has one final frontier to cross: Earth's atmosphere. Our atmosphere is a layer of atoms and molecules several hundred kilometres thick, quite tenuous at the top, where it fades into space, but fairly dense at the ground, where we can breathe. It is not a static medium, either; it churns with turbulence and currents introduced by temperature and pressure gradients. You don't need me to tell you that, of course – just go outside on a windy day. We already know that the sky scatters sunlight, but what other effects does it have on photons hitting the atmosphere from space?

Well, first we should note that the atmosphere is not transparent to all light: high-energy ultraviolet and X-ray photons are readily absorbed by molecules in the air and never make it to the ground. For example, ozone (a molecule composed of three oxygen atoms bound together) in the stratosphere is particularly good at absorbing ultraviolet light with a wavelength of about 200 to 300 nanometres, also known as 'UV-B' radiation. Interestingly, ultraviolet light is also responsible for helping to generate ozone: a photon of the right energy can split a regular oxygen molecule (two oxygen atoms bound together) and each of the constituent atoms can go on to bind with other oxygen molecules to form ozone.

Although the peak of the solar spectrum is in the visible part of the spectrum, it still emits a considerable amount of ultraviolet light. The absorption of solar ultraviolet radiation by ozone is tremendously beneficial to biological life because high-energy photons can have a damaging effect on cells, as anyone who has had a sunburn will attest. More seriously, high-energy photons can cause genetic damage by colliding with, and potentially breaking, DNA molecules in our cells. Incorrect repairs of the damaged DNA can introduce biological mutations that might result in the growth of cancers. This was why the discovery of a hole in the ozone layer in the 1980s was such a global concern; it was a breach of Earth's natural shield to ultraviolet radiation. The problem has been somewhat rectified in recent years with the success of the Montreal Protocol, signed in 1987 to limit the use of aerosols and other substances that were found to deplete the ozone layer.

At longer, infrared wavelengths, to the red side of the visible part of the electromagnetic spectrum, the atmosphere is almost opaque. In the near-infrared bands, at wavelengths of a few microns, there are a few 'windows' in the transparency of the atmosphere. Photons with energies corresponding to these wavelength ranges make it through. But the vast majority of infrared radiation is readily absorbed by water molecules in the atmosphere.

As we go beyond the infrared part of the spectrum and move to lower energies, the atmosphere becomes transparent again. For electromagnetic radiation with a wavelength of about 1 millimetre or more, we transition from the infrared part of the spectrum into the microwave and radio bands. Most of these photons can pass through the atmosphere quite easily.

The fact that the Earth's atmosphere blocks out large swathes of the electromagnetic spectrum is one of the main reasons for putting telescopes, especially those operating in the ultraviolet, X-ray and infrared parts of the spectrum, into space: simply, to catch those photons *before* they are lost to our pesky atmosphere. Putting telescopes above the atmosphere also helps overcome another problem: the degradation of the sharpness of images we can take from the ground.

Travelling across the vacuum of free space, photons travel at the speed of light, a fundamental upper limit. But as we know, as soon as a photon encounters a medium – anything other than a vacuum – its speed of propagation changes. It slows. We know this phenomenon as refraction, which is a change in direction of a wave, or a photon, due to a change in speed. So all light travelling from space into our atmosphere is refracted.

Unlike the case of a submerged straw, the situation with the atmosphere is a bit more complex: the composition, density, pressure and temperature of the medium is changing all the time. We can approximate the atmosphere as a collection of different 'cells' that each have their own impact on the propagation of the photons. As the light passes down through the atmosphere, it encounters lots of little cells of atmosphere with different refractive indices. By the time a single extra-terrestrial photon reaches our telescope, it is likely to have been knocked off its original path a little bit.

What effect does this have? Simple: it slightly blurs any image we can make using a telescope on the surface of the Earth, so the level of detail we can discern in objects such as nebulae and galaxies is reduced. We call this effect 'seeing'. The poorer the seeing, the more blurred the image. This problem gets worse the more atmosphere you have to look through, and this is why the best telescopes in the world are placed at altitude, preferably on high, dry mountains or plateaus; to look through as thin a layer of atmosphere as possible.

The Hubble Space Telescope produces such extraordinarily sharp and detailed images not because it is a large telescope, but because it collects light that has not passed through the atmosphere. Compared to the best ground-based optical telescopes, Hubble is quite small: its collecting mirror is just 1.5 metres across, which is tiny compared to the 8-metre Very Large Telescopes or 10-metre Keck telescopes. It is easier to build large collecting mirrors on the ground, and they are simply too big and heavy to be put into space at any reasonable cost. The largest ground-based optical telescopes that currently exist will soon be dwarfed by a new generation of 30- to 50-metre-wide optical telescopes, one of which is the Extremely Large Telescope that is now under construction in the Chilean Atacama.

Regardless of the size of a telescope, unless the distorting effects of Earth's atmosphere are compensated for, the sharpness, or resolution, of images that can be taken by ground-based telescopes operating in the visible and near-infrared bands will always be fundamentally limited by the seeing. For that reason, some telescopes, including the new Extremely Large Telescope, have a remarkable technology called 'adaptive optics' which can actively deform the surface of the collecting mirror, responding to, and cancelling out, the aberrations introduced to the paths of photons as they pass down through the sky. This can restore the crispness of the image as if the atmosphere wasn't there, rivalling what can be achieved by the likes of the Hubble.

The bigger the telescope, the more photons it can collect and therefore the further it can see. Remember, the flow of energy, or 'flux', arriving at Earth's surface from other stars and galaxies scales with area: you can think of the number of photons passing through a sheet of paper every second. Make that sheet, or the telescope's mirror, bigger and the more photons you will capture. And that is the main objective of any telescope: to intercept photons at the end of their long journey across the Universe.

Journey's end for a photon is the actual process of astronomical observation. The photon must be recorded by a detector and turned into a digital signal. Humans can 'detect' photons because we have sets of millions of specially modified neurons at the back of our eyes called photoreceptors, honed over millions of years of evolution. The photoreceptors can absorb photons – again, ultimately a quantum mechanical effect – and cause the cell to respond by sending a signal that is transmitted to the brain via the optic nerve. Unfortunately, our eyes don't have a large collecting area and cannot take the long exposures needed to see very faint objects, but our brains arguably do have a reasonable, if sometimes unreliable, storage system: memories.

In the very early days of observational astronomy there was no easy method to record the light from observations: you had to look through the eyepiece and sketch what you saw. Although telescopes aided the eye in magnifying distant objects, long exposures were still not possible. The development of photography was a revolution for the field because light could actually be recorded accurately, and photographic plates and film could be exposed for extended periods, collecting light over minutes or hours and producing the first truly deep maps of the cosmos.

Photographic plates were scoured for distant, faint objects, and this was the method Hubble used to detect the presence of Cepheid variable stars in Andromeda. There was an elegance and craft to this method no doubt, but the era of photographic detection, at least in the traditional sense of silver emulsions, glass plates and so on, has long since passed.

Since their introduction in the mid-1970s, the bread and butter of hardware for astronomical detectors in the visible part of the electromagnetic spectrum has been the charge-coupled device, or CCD. The same technology is in your digital camera and smartphone. The CCD is the digital equivalent of the photographic plate but made up of a grid of tiny detectors that represent picture elements, or as we now call them, pixels.

James Janesick and Morley Blouke, pioneers of CCD technology, suggested a simple way of thinking about how a CCD works. Imagine that each pixel is like a bucket on a conveyor belt, with a series of rows of belts making up the full array. When it rains, these buckets can start to fill with water, and if the rain falls harder in one spot, then those buckets will fill with more water. After some length of time the conveyor belts can carry the buckets on each row to the edge of the array, where the amount of water in each bucket can be poured into a measuring vessel and recorded.

If we are interested in a rain of photons rather than drops of water, we need something other than a bucket to do the collecting, and to record the information digitally we need to convert the light signal into an electric charge. This is where the main component of the CCD comes in: a semiconductor. A semiconductor is a type of material that conducts electricity, but only under certain conditions. If those conditions can be controlled, then it is possible to manipulate electric charge.

Crystalline silicon is a semiconductor, and we can artificially enhance its properties through a process called 'doping'. Doping

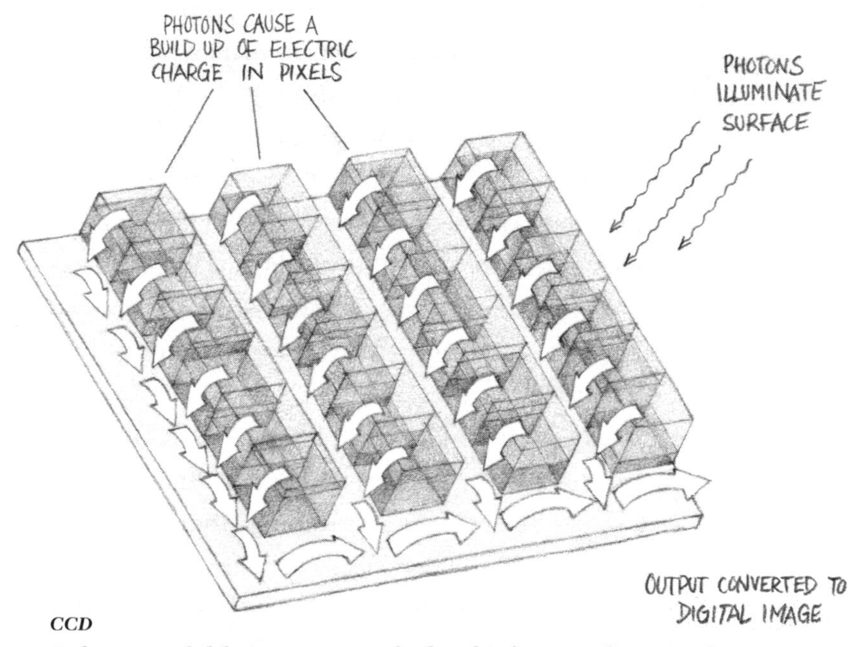

CCD

A charge-coupled device, or CCD, can be thought of as a two-dimensional array of receptacles that 'fill up' with electric charge as photons 'rain down' on them.

is where additional atomic elements are deliberately introduced to contaminate the silicon crystal lattice in such a way that the 'impurities' affect the conducting properties. Each pixel in a CCD contains a little crystal of doped silicon. When a photon of the right energy hits the silicon, the semiconductor can absorb the photon. This transfers the photon energy to an electron, 'promoting' it into part of the semiconductor structure called the conduction band. This is called the 'photoelectric effect' and its description is what Einstein was awarded his Nobel Prize for in 1921. Under normal circumstances this electron would release the extra kick of energy and return to its original level quite quickly. But if a voltage is applied to the semiconductor, then the electron can be artificially held in the conduction band for a while.

As more photons hit the detector, further electrons can be promoted in this way, and since electrons have individual negative charge, a bulk negative charge builds up in each pixel. The size of this charge is proportional to the amount of light – the number of photons – hitting the detector. The CCD can be exposed to light for some time to build up an appreciable signal, or up to a maximum point when the charge saturates (the bucket overflows). Once the exposure is finished, the final task is to 'read out' the CCD and turn the charge recorded by each pixel into numerical values in an image.

Read-out is achieved though clever electronic structuring of the CCD array that allows the charges in neighbouring pixels to be shuffled along to the edge of the array through the appropriate application of voltages. The charge from every pixel is then sent through a device called an analogue-to-digital converter. This changes the signal into a discrete range of numbers that can be read by a computer. When those numbers are arranged in the same pixel grid pattern as the CCD, we have an image. Although the exact mode of operation of CCDs vary from chip to chip, and are slightly different depending on the wavelength of the light being observed, this general process is the basis of nearly every astronomical image you see.

The most distant starlight we have detected so far – actually the combination of billions of stars – left its galaxy when the Universe was less than five hundred million years old. Some of those photons happened to hit the mirror of the Hubble Space Telescope as it stared for weeks at a single patch of sky. Reflected off the mirror and focused, the photons – now billions of years old – created a small signal in one of its CCD cameras. We see that distant galaxy as a tiny, almost indistinct clutch of pixels just slightly brighter than those around it – they are the whisper of billions of stars carried over space and time.

And that is the fate of some starlight: to hit a telescope mirror and to be reflected and focused onto a camera, finally depositing its energy after what may have been billions of years of flight. That we can actually record those whispers of energy and, more importantly, store that information, is what allows us to explore and understand the parts of the Universe we can never visit. That is the craft of astronomy.

FOUR

DARK ENERGY'S IMPRINT

I was once sat in a cosmology seminar that opened with the line, 'What is the Integrated Sachs–Wolfe Effect?' This prompted a few moments of predictably awkward silence from the audience, until the speaker continued, 'It's like the Sachs–Wolfe Effect . . . but integrated!' As cosmology jokes go, I thought this was one of the better ones. Okay, maybe you had to be there.

Rainer Sachs (1932–) and Arthur Wolfe (1939–2014) were two astrophysicists who in 1967 wrote a paper in the *Astrophysical Journal* describing what happens to cosmic microwave background photons as they leave the Surface of Last Scattering. I want to tell you about the Integrated Sachs–Wolfe Effect because I think it's a beautiful observational illustration of how the very evolution of the Universe, in particular the imprint of the mysterious dark energy, can be written in the light that shines through it.

We have already talked about how photons travelling across the Universe have been subjected to redshift, and that this redshift arises from a combination of effects. One is the expansion of space between the time of the emission of the photons and the time we observe them, called the cosmological redshift. The second is because of the Doppler shifting of the frequency of a photon emitted by a source that is travelling with some relative velocity towards or away from us, perhaps due to the rotation of a galaxy or the 'peculiar' motion caused by local gravitational attraction with other galaxies.

There is a third source of redshift that we haven't talked about which is central to this tale, and it's called gravitational redshift. Its origin lies in a theory called general relativity, one of Einstein's most important contributions to our understanding of nature. So, before we get much further we should delve a little deeper into the story of how light travels through the Universe, or rather, how light travels through space and time. We need to understand a bit about general relativity.

We use the idea of 'potential wells' to describe the strength of the gravitational field around a massive object. The word 'well' alludes to the notion that a gravitating mass will deform the fabric of spacetime – a bowling ball on a rubber sheet is a common analogy – such that other bodies can fall into, or be deflected by, these depressions, like a marble rolling along the rubber sheet.

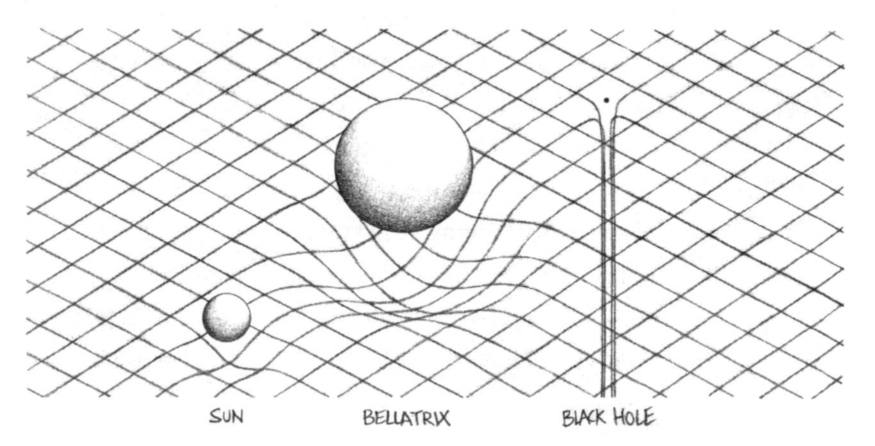

Curvature of spacetime

General relativity describes gravity in terms of the curvature of spacetime. Spacetime has four dimensions, but for illustration it is shown here as a two-dimensional sheet. The presence of mass or energy density – like a star – deforms the sheet. Photons travel along paths in spacetime, so their trajectories can be deflected as they pass near massive objects. Black holes cause such extreme curvature that any photons within a certain critical distance from them cannot escape.

But what does it mean for there to be a 'fabric' of spacetime, as it is often described? What is spacetime, and why are space and time linked at all?

The straightforward answer to the question sounds obvious when you write it down: all events in the Universe occur at some location in regular three-dimensional space (somewhere) and also at some point in time (sometime). So any event in the past, present and future Universe can be given in terms of four values: x, y, z and t, measured relative to some reference. We call that reference the 'reference frame', and as we shall see, this is a fundamental concept in relativity.

We can define any frame of reference we wish to, and all the laws of physics can be written down in terms of the spacetime coordinates in that frame. But we can go further. An 'inertial' frame is one which has no external influences: the physical laws within it are absolute, and you would measure exactly the same laws in another inertial frame. A practical example of an inertial frame is a car that is travelling with constant velocity. Within the frame of the car, you don't feel like you are moving at great speed – you are 'at rest' with respect to the car around you and its contents. Drop something and it falls to the floor through gravity. But if the car starts accelerating, it becomes a non-inertial frame. You feel a force on your body, pushing you back in your seat as the car speeds up. Within your frame, if you had no knowledge that you were in a car, it is as if this force pushing you back suddenly appears out of nowhere. You wouldn't necessarily be able to determine whether it was caused by an external influence on your frame or something internal to it. We sometimes call that a 'fictitious force'.

You can imagine multiple inertial frames with different coordinate systems, and have all these frames moving with constant velocity with respect to each other. We can convert physical measurements in one frame to those in another through simple 'coordinate transformations'. Imagine standing on a platform, watching a train go by. Let's say the train is moving with a constant speed of 50 miles per hour. Within the train someone is walking down the aisle in the same direction as the train is travelling, and you can see them through the window. For argument's sake, let's also say their walking pace is about 1 mile per hour. You, watching from the platform, would measure the speed of the person inside the train as 51 miles per hour: their walking speed plus the train's speed. But within the train, someone sitting on a seat would clock the fellow passenger walking down the aisle at just 1 mile per hour, because that observer is in the same frame as the walking passenger.

The platform and the train are two inertial frames, and we could easily 'correct' our platform measurement of 51 miles per hour by changing our coordinate system to one that is moving with the same velocity as the train. This is called Galilean relativity. Einstein refined this concept into his 'special' relativity and eventually into general relativity. Galilean relativity works fine for many situations, but special relativity must be considered whenever the velocities of objects approach the speed of light. One of the fundamental postulates of special relativity is that the spacetime interval – the difference in coordinates – between two events is independent of the inertial frame those events are measured in. This is an outcome of the constancy of the speed of light, which all observers measure at the same value, regardless of their frame of reference.

What this means is, if the person walking down the aisle of the train is replaced with a photon travelling at the speed of light, then the observer on the platform would not measure the speed of that photon as 'the speed of light plus 50 miles per hour', but just the regular speed of light. The added speed of the train makes no difference. The passenger sitting on the train would also measure the photon passing them down the aisle at the speed of light.

Where does spacetime fit in to all this? The set of coordinates x, y, z and t essentially are spacetime. As human observers we perceive the 'somewhere' simultaneously and can move about at will in three-dimensional space. As such, we have a pretty good intuitive understanding of what three-dimensional space actually is. But our perception and experience of time is different. We cannot travel arbitrarily through time, we simply have to go with the flow, which as far as we can tell only flows forward, at least on macroscopic scales. And that is why the concept of our existence in a four-dimensional spacetime can be quite hard to grasp.

Nevertheless, all events in the Universe do occur somewhere and sometime, and we can think of the spacetime interval between two events within the four-dimensional construct. An equation that describes this interval in a given frame – which is basically the difference between any two events' spatial and temporal coordinates – is called the 'metric'. The exact form of the metric is fundamental to our description of how objects behave in spacetime, and – importantly – it can account for spacetime that is curved. That is the crux of general relativity.

We don't have time to go through the whole story of general relativity here, but let's explore some of the key concepts, as it will help us understand the behaviour of light as it travels through the Universe. First let's think about mass. Newton taught us that a force acting on a body will cause that body to accelerate, and the magnitude of the acceleration is equal to the ratio of the force to the mass of the body. Specifically, we think about the 'inertial' mass of the body in that case.

Newton also taught us about gravity, showing that there is an attractive force between two bodies that is proportional to the product of their masses and inversely proportional to the square of the distance between them. Let's consider two objects: a tennis ball as one body and the Earth as the other. Obviously, the mass of the Earth dwarfs that of the tennis ball, and so it dominates the gravitational force in this example. We can write an equation to describe the force acting on the tennis ball due to the gravitational field of the Earth, which we describe as the gradient, or slope, of the gravitational potential. In this case we think about the 'gravitational mass' of the tennis ball – the mass that 'feels' the gravitational force due to the Earth.

But the gravitational force on the tennis ball acts like any other force: it produces an acceleration. Release a tennis ball and it will accelerate towards the ground. But, as we have just learned, that acceleration is related to the ball's inertial mass. It is therefore possible to write down an equation relating the 'inertial' mass to the 'gravitational' mass of the tennis ball, which in words goes like this: the tennis ball's *inertial* mass multiplied by its acceleration is equal to the ball's *gravitational* mass multiplied by the gradient of the Earth's gravitational potential. This is called an equation of motion.

If inertial mass and gravitational mass are equivalent – and it's not immediately obvious that they should be – then the equation of motion implies that the acceleration of the tennis ball due to the gravity of the Earth is *independent* of the mass of the ball, since the tennis ball's mass appears as a term on both sides of the equation. You could make the ball ten times more massive and the acceleration would be the same, because the inertial mass and gravitational mass 'cancel out' in the equation of motion.

The observational outcome is that all objects, no matter what their mass, will fall to the ground with the same acceleration due to gravity, near the Earth's surface.

To many people this sounds counterintuitive, at least initially. The gut reaction is to assume that heavy objects fall faster than light ones. But this is demonstrably not true – most famously shown by Galileo in his apocryphal experiment of dropping two objects of different mass from the top of the Tower of Pisa: they fell at the same rate towards the ground. A more recent, convincing demonstration was conducted during an Apollo 15 Moon walk. While on the lunar surface, the astronaut David Scott held up a falcon feather and a hammer in front of a television camera and then dropped them simultaneously, clearly showing that they hit the lunar soil at the same time.

That demonstration doesn't work so well on Earth because falling objects experience another force that opposes gravity: air resistance. This is more effective for the feather, and so when the same experiment is done on Earth, the hammer indeed falls to the ground faster, while the feather lags behind as it gently floats down. Nevertheless, this 'equivalence principle' between gravitational and inertial mass is one of the fundamental roots of relativity and basically states that a gravitational field is equivalent to an acceleration.

We can explore this further with a thought experiment, a favourite tool of Einstein. Imagine an elevator in free fall. Inside the elevator is a person. From the outside we can of course observe that the elevator and its contents are in free fall, plummeting towards the ground from some height due to the Earth's gravity. The equation of motion is straightforward: the elevator and the occupant are accelerating directly towards the Earth at a constant rate, increasing in velocity by about 10 metres per second every second, a value known in physics as g, or the acceleration due to gravity near the Earth's surface.

Inside the elevator the occupant is suspended in mid-air, as if free from the effects of gravity, because the occupant and the elevator are falling at the same rate. This is why astronauts on board the International Space Station appear to be weightless. It is not that the Earth's gravity is not acting on them, it is because both they and the station are free falling towards the Earth. In the case of the International Space Station (and all satellites), the craft is also travelling with some velocity tangentially to the Earth's surface. So although the station is 'falling', it is also moving 'forward'. That combination is tuned such that, although the station is always falling towards the centre of the Earth, it never reaches the ground because of the curved surface of the planet: the altitude remains stable. This is an orbit.

We can describe the point of view of an external observer watching the falling elevator and the observations of the oblivious occupant inside it as different frames of reference. Inside the elevator we have transformed to a frame of reference that is accelerating at g, and the local effects of gravity – such as being held to the floor of the elevator – are 'transformed away' for that observer.

We could turn things around. Put the elevator somewhere out in space, far away from any large gravitating mass such as a planet. We'll now call it a space capsule. If the capsule is at rest or travelling with a constant velocity (that is, not accelerating) through empty space, then the occupant really will be weightless because there is no gravitational field. But if that capsule started accelerating, then the occupant would experience the same fictitious force that pushes you back in your seat in an accelerating car. If the acceleration of the capsule was equivalent to g and kept constant, then the occupant of the capsule would be held to the 'floor' in exactly the same manner that you are held to the ground. This is the *Weak* Equivalence Principle, which states that an accelerating frame (like the capsule) is equivalent to a static frame in a gravitational field (like standing on the surface of the Earth).

You will have noticed that inside the accelerating capsule we have simulated gravity, so the occupant could walk around, jump up and down, and so on, in much the same way they would be able to on Earth. Unfortunately, maintaining a constant straight-line acceleration like this is unfeasible – it would take too much fuel and the craft would have to be constantly moving, increasing in velocity all the while. An alternative is to create a cylindrical capsule and spin it around its long axis. In this case there is still a force exerted due to the movement of the capsule, but it is a centrifugal force, causing the occupant to be pushed against the curved surface of the craft. Get the spin just right, and you can mimic Earth's gravity within the capsule. This is a concept much used in science fiction, most famously in Stanley Kubrick and Arthur C. Clarke's *2001: A Space Odyssey* (1968).

Now, the *Strong* Equivalence Principle is where general relativity really starts to kick in. It states that every natural law of physics is the same in a reference frame that is freely falling in a gravitational field as it would be in the absence of that field, regardless of the strength of the field. This can lead to some interesting effects.

Imagine that our falling elevator contained a device to shoot a light beam from one side of the elevator to the same point on the opposite side – a perfectly horizontal beam. During free fall, the observer inside the elevator would see exactly that. But for an external observer carefully watching the beam (let's say the elevator has a glass side so we can see through), the light would

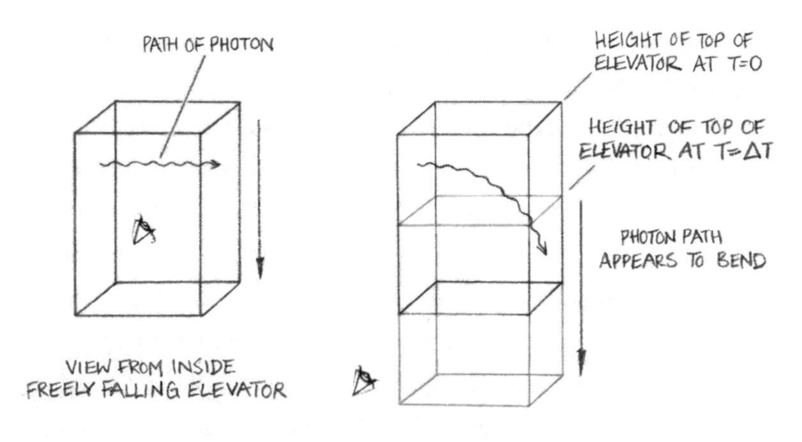

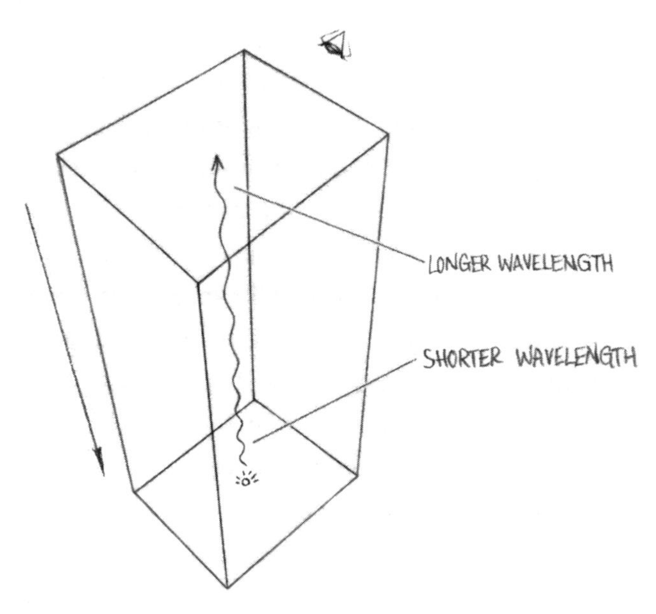

Gravitational lensing and gravitational redshift

An elevator in free fall, as viewed by two observers – one falling with the elevator and one at rest, watching from the outside – can be used to describe some of the fundamental concepts of relativity, including the phenomena of gravitational lensing and gravitational redshift of light.

be emitted from one side and then appear to arc downwards towards the floor.

Remember that the speed of light is measured to be the same inside the elevator as it is measured from the external frame of reference. From the point of view of the external observer the elevator has fallen towards the Earth *during* the time it takes for the light to travel from wall to wall. So the physical point on the opposite side of the elevator where the light beam is intercepted is now lower (relative to the ground) than it was when the beam was emitted on the other side. This means that the external observer sees a light beam that appears to bend as it travels through Earth's gravitational field. This is the basis of the gravitational deflection, or lensing, of light by gravity, and is one of the key predictions of general relativity.

Observational evidence of the gravitational lensing of light was first demonstrated in 1919, not long after Einstein first described general relativity. Arthur Eddington (1882–1944) travelled to the island of Principe, to the west of Equatorial Guinea, to observe a total solar eclipse. During the darkness of totality, Eddington was able to measure the position of a background star near the Sun's edge. When that position was compared to one obtained in an observation of the same star, but when the Sun's disc was not in front of the star field, a difference in position was measured. That difference was the predicted deflection of light from the background star as it passed through the Sun's gravitational field. The background starlight had been gravitationally lensed.

Now we get to gravitational redshift, which we can also explain with the free-falling elevator. Instead of a light beam crossing the elevator from side to side, we can put the light source on the floor, so that the beam is going in a vertical direction. Let's make the light source a bit like a clock, so that it emits pulses of light at a rate exactly tuned to the frequency of the electromagnetic radiation. Inside the elevator, the observer would measure the time it takes for each pulse of light to travel from the floor to the ceiling as the height of the elevator divided by the speed of light. The frequency at which pulses arrive at the ceiling will be identical to the rate at which they leave the source on the floor.

Now let us switch to our external observer. Imagine them on top of a tall tower, looking down as the elevator falls past them towards the ground. This time we'll make the ceiling transparent so we can see the upcoming pulses of light. Now, from this point of view, the source of the light on the elevator floor is falling away from the external observer with increasing velocity. Like a receding siren, this motion introduces a Doppler shift, decreasing the light's frequency. So the external observer looking down sees a redshifting of the light pulses. But due to the equivalence between an accelerating frame and a static frame in a gravitational field, the light signal travelling upwards in the free-falling elevator is just like a photon climbing out of a gravitational potential well. This means that any photon travelling out of a gravitational potential well towards an external observer will be redshifted, and this is the gravitational redshift. The opposite will be true for a photon travelling into a potential well – it will be gravitationally blueshifted.

A related effect is that, along with the decrease in the frequency of the electromagnetic radiation, the rate at which the external observer counts the emission of light pulses coming from the floor of the elevator also decreases as it falls. The ticking of the clock slows. One of the most disconcerting outcomes of general relativity is that clocks run slower when they are deeper in a gravitational potential well.

So, if you had two synchronized clocks and left one at sea level and took the other up to the summit of Mount Everest, and then left both clocks alone for a while, when you return you will find that the clock on top of the mountain has elapsed more time than the one on the ground. The amount that they differ depends on the difference in gravitational potential energy between the summit and ground (the mountaintop is, of course, higher up in Earth's gravitational field). And this does not just affect time-keeping: anyone staying with the clock on the mountain will age faster than those who remain on the ground.

This all sounds like a very strange way for nature to behave, and you would be forgiven for thinking that these 'thought' experiments just make weird predictions that can't actually happen in reality. Can it really be true that people living in Tibet age faster than those in the Netherlands? Fascinatingly, both gravitational redshift and gravitational time 'dilation' have been convincingly experimentally measured on Earth. The Pound–Rebka experiment, as it became known, was conducted by Robert Pound and Glen Rebka at Harvard University in 1959. They demonstrated that gamma rays emitted by a sample of a particular isotope of iron (iron-57, to be exact) located at the top of a building underwent a gravitational blueshift when the photons were received in the basement, about 23 metres below.

Another test, called the Hafele–Keating experiment, conducted by Joseph Hafele and Richard Keating in the early 1970s, put atomic (highly accurate) clocks on aircraft which were then flown around the world. The airborne clocks were initially synchronized with a reference clock that remained grounded, at rest in the laboratory during the flights. General relativity predicted a difference in the time between the clocks at altitude and at ground level of about one hundred and fifty billionths of a second. There is actually another time dilation effect predicted by the special theory of relativity that accounts for the fact that the aircraft are also moving with appreciable velocity relative to the ground. When the flights landed and clocks were compared, the ground-based clock showed a difference in excellent agreement with that expected from theory. There have been many other similar tests since, all confirming Einstein's prediction with increasing accuracy.

Although this seems like a negligible effect for clocks in the vicinity of Earth (who cares about a few billionths of a second here and there?), the fact that clocks run slower on the surface of the planet compared to those at altitude certainly needs to be taken into account in, for example, global positioning satellites, which have orbits with a radius (altitude) of some 20,000 kilometres. The satellites must keep their time to an accuracy of a few tens of billionths of a second relative to those on Earth if we are to use them to measure our latitude and longitude with positional accuracies of about 10 metres or better. If we know the altitude and orbital velocity of each satellite, the time gain experienced by the satellites can be corrected because we know about general relativity. And if we fail to account for time dilation, the satellite clocks would start gaining time at a rate of tens of microseconds per day, relative to clocks on Earth. This would very quickly render the accuracy of the global positioning system no better than 10 kilometres, getting worse all the time.

Einstein's achievement in general relativity was to mathematically describe all this physical behaviour by considering how mass and energy density deforms spacetime and the impact this has on the passage of light through it, as measured by different observers. His theory completely revised our picture of how gravity works. The mathematical expression for this relationship between time and space and the density of matter and energy are encapsulated in Einstein's 'field equations'.

Implicit in the field equations is a geometric description of spacetime as a 'manifold'. This is just some mathematical jargon used in geometry. The surface of a sphere is a type of two-dimensional manifold, or 2-manifold, as is the surface of a cylinder or a torus. These 2-manifolds can exist in a three-dimensional space, like the surface of the Earth. A solid ball – the surface plus the interior – is a type of 3-manifold. Spacetime is a 4-manifold. If you want to think of a 'fabric' of spacetime, then this manifold is it.

General relativity describes gravity in terms of the curvature of the manifold. The presence of matter or energy deforms spacetime, and this deformation is what causes matter to move due to gravity. Particles, like photons, travel in curved spacetime following 'geodesics', which are the shortest distance between two points on the manifold. Just as the shortest distance between two points on the 2-manifold of the surface of the Earth are arcs of 'great circles', the paths of photons through the matter- and energy-filled Universe follow the curvature of spacetime.

We can think about the gravitational landscape of the Universe at the time of recombination. Imagine regions where the density of matter was slightly higher than average, and where it was slightly lower than average. These imperfections eventually grew into the large-scale structure of matter in the Universe. You can think of the broad over- and under-densities of this primordial landscape in terms of how they deform the spacetime of the early Universe like a range of hills and valleys.

Sachs and Wolfe predicted that the photons escaping from the hills and valleys at the point of recombination, when they were no longer trapped in the photon-baryon fluid, would be subject to gravitational redshift. The photons in the denser regions would have to lose a bit of energy to climb out of their valley compared

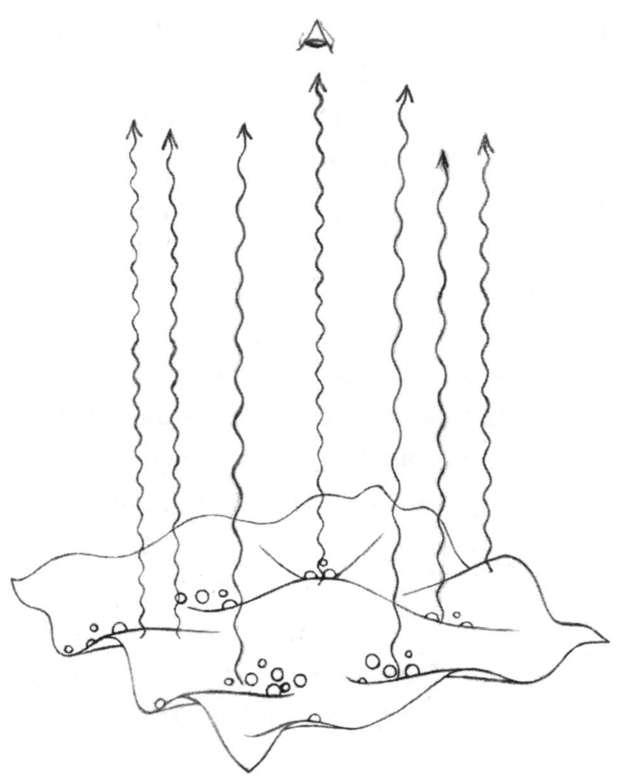

Sachs–Wolfe Effect

The Sachs–Wolfe Effect: photons escaping from regions of higher density at the Surface of Last Scattering undergo gravitational redshift as they 'climb out' of gravitational potential wells.

to those at the tops of hills – the low-density regions. So, at the point they were emitted, the cosmic microwave background photons were subject to gravitational redshift to varying degrees. That's the regular Sachs–Wolfe Effect, and it creeps into the power spectrum of temperature anisotropies measured by experiments like the Planck satellite as temperature fluctuations on the largest scales we can measure, where this effect is most pronounced.

As we have seen, the journey of a cosmic microwave background photon can be anything but uneventful. They are the photons that have the longest path to our detectors and must traverse nearly the entire observable Universe to reach us. That means that during their over thirteen-billion-year-long journeys, not only are they subject to the cosmological redshift due to the expansion of the Universe, they must also travel through large-scale cosmic structures that are themselves evolving over time.

Imagine that you could speed up time for everything around you, while time for you personally ran at a regular pace. Let's say that in one second of your time, one hundred years passed in the world around you. Now take a long walk. On your journey you see monumental events unfold: towns and cities rise and fall. The landscape changes. This is a bit like what it's like for a cosmic microwave background photon: the gravitational landscape of the Universe is changing during its journey as cosmic structures grow.

A cosmic microwave background photon racing through the Universe will traverse gravitational potentials that have grown, and are still growing, during the journey. This can introduce a gravitational redshift on the light. Imagine a photon travelling towards a large mass. That large mass causes a great depression in spacetime, as described by the field equations. Like a ball rolling down a hill, on the inward trajectory towards the mass the photon gets blueshifted, gaining in energy. As the photon shoots past the mass and is on the outward trajectory, it is like the ball is climbing back up the hill. As such, the photon must lose energy to climb out of the potential well, redshifting it. There may be many such encounters on the journey, as the photon passes through different cosmic structures, and the culmination of this is called the Integrated Sachs–Wolfe Effect. Yep, it's like the Sachs–Wolfe Effect, but integrated.

Now, if the amount of energy gained on the inward trajectory is the same as that lost on the outward trajectory, this means that the depression in spacetime is symmetric. The blueshift and redshift cancel each other out. So when we detect the photon as it emerges from the potential well, we won't necessarily be able to distinguish it from a photon that had not passed through the same structure. But what if this is not the case? There are two possibilities: either we observe a net blueshifting or a net redshifting of the light as it travels through a gravitational potential.

To be blueshifted, the photon must expend less energy climbing out of the potential well than it gained on its way in, and vice versa. One way for this to happen is through the evolution of spacetime on large scales due to the expansion of the Universe. Cosmic expansion could change the depth of the potential – the amount of curvature of spacetime – in the time it takes for the photon to cross a large-scale structure. It follows that, beyond the regular cosmological redshift experienced by all photons, the detailed history of cosmic expansion could be imprinted on cosmic microwave background photons through the Integrated Sachs–Wolfe Effect. If so, we have another powerful observational tool to refine our cosmological model of the Universe.

In order to find out how fast the Universe is expanding, you need some method to compare physical scales at different cosmological redshifts. It is easy to start substituting the word 'redshift' for distance when we talk about astronomical objects such as galaxies, but it is important to remember that redshift really just reflects the ratio of the 'scale factor' between the present day and some earlier time. The scale factor describes the evolution of the expansion of the Universe since the Big Bang and is simply defined as the ratio of the 'proper' distance between two objects measured at two different times.

Proper distance is defined as the spacetime interval between two events with the time coordinates fixed. Consider two stars, say the Sun and its nearest neighbour, Proxima Centauri, about four light years away. If you could run a tape measure from the Sun to Proxima Centauri, in the time it took you to travel from one star to the other the stars may have moved with respect to each other. So the actual distance you read off your tape when you arrive at Proxima Centauri depends on the time difference too. If you travelled slower or faster, taking a longer or a shorter time to get there, you might read a different distance. So, that's not a 'proper' distance. Instead, if you could travel instantaneously from the Sun to Proxima Centauri, no time will have passed and so the difference in the spacetime interval between these 'events' is simply due to the stars' instantaneous separation in regular three-dimensional space. That's a proper distance.

Now scale things up a bit: imagine the proper distance between two widely separated galaxies, so far separated from one another that their mutual gravitational attraction is negligibly weak. Imagine we could travel instantaneously between the galaxies to measure the proper distance separating them. Do that once, then wait for a few billion years and do it again. The expansion of the Universe – that is, the expansion of space – in the intervening years will have increased the proper distance between the two galaxies. If we could repeat this for the same pair of galaxies observed at all cosmic times, we could very accurately measure the expansion history of the Universe.

The cosmological redshift of light from distant galaxies can be related to our tape measurement problem: a photon leaving a distant galaxy takes a finite time to reach us, and in the intervening year's the Universe has expanded. By the time we detect that light, it has undergone a redshift that depends on the size of the Universe at the point of emission and the size of the Universe today. The problem is that if we don't know *how* the Universe has expanded in the time it has taken the light to reach us, it is difficult to translate a measured redshift into a distance. Different cosmological models predict different relationships between redshift and the evolution of the scale factor, and that is why determining this relationship empirically is a key part of our understanding of cosmology. But it's hard.

We have already learned how Edwin Hubble first showed that the Universe was expanding: he plotted the recession velocity of galaxies as measured by Vesto Slipher, versus Hubble's estimates of their actual distance. This graph became known as the Hubble Diagram. The slope of the correlation between recession velocity and distance is called the Hubble Constant and gives the rate of expansion of the Universe per unit distance. That value is measured to be about 70 kilometres per second per million parsecs (one parsec, the unit of distance astronomers tend to use, is just over three light years), although there remains a bit of contention about its exact value even today.

Hubble's original data suggested a constant rate of expansion, but in the 1920s telescopes were not powerful enough, and instrumentation was lacking in sophistication, to see much further than the local Universe. Over the century since, astronomers have sought to address this issue by building ever larger telescopes and sensitive cameras and spectrographs. We can now measure the redshifts of galaxies seen across the bulk of cosmic history.

If you can collect enough photons, redshifts are very easy to measure: first, disperse the light of a distant object into a spectrum, telling us how many photons are being emitted at different energies. Then, with a good knowledge of the energies of various atomic transitions that occur in different elements, just compare the observed wavelength of the spectral emission features to what they would be if measured 'at rest' on Earth. For example, galaxies that are forming new stars tend to have lots of ionized hydrogen. Newly born massive, luminous stars irradiate their birth gas clouds with ultraviolet light. This energizes the atoms in the gas, promoting electrons to higher energy levels or removing electrons from atoms completely. When these electrons eventually return to their original energy level, they emit a photon with a wavelength corresponding exactly to the energy of the atomic transition.

A prominent, bright emission line is called hydrogen-alpha, which occurs at exactly 656.28 nanometres, appearing as a pronounced spike in a star-forming galaxy's spectrum. Any observer in a frame of reference that is at rest with respect to a cloud of gas emitting hydrogen-alpha light would measure the same wavelength. But for observations of a distant galaxy this emission line might be redshifted to, say, 1,968.84 nanometres by the time we detect it in our frame of reference, which of course is not at rest with respect to the distant galaxy, since it is receding away with cosmic expansion. By comparing the observed wavelength of hydrogen-alpha, or some other spectral feature, to the 'rest frame' value, we get the redshift of the source. We can design observational surveys to measure the redshifts of thousands of galaxies this way. These 'redshift surveys' give us some impression of the distribution of galaxies in space and, by comparing their properties at different redshifts, how they have changed over time.

But remember that the redshift only really tells us about the relative size of the Universe between the time of emission and the time of detection. What we would like to measure is the actual distance to the object. This is far trickier.

How do you actually know the distance to something? We have already encountered a 'standard ruler' in baryonic acoustic oscillations: if we can measure the size of the ruler at different redshifts we can track cosmic expansion. We have also talked about 'standard candles', objects of known luminosity. By comparing the known intrinsic luminosity to the candle's observed brightness, which dims according to the square of the distance, we can figure out how far away it is. Standard candles were fundamental to Hubble's original discovery of the expanding Universe in the 1920s, and they are also fundamental to the story of the discovery of the *accelerating* expansion of the Universe three-quarters of a century later.

Unfortunately, there are actually very few types of object that are reliable standard candles. They must be sources of electromagnetic radiation with a very well-defined luminosity (the amount of energy they emit every second), with little spread in that luminosity across different objects. Cepheid variable stars have an average energy output that is correlated with their pulsation period, so we can use that periodicity as a proxy for their luminosity. Unfortunately, if you want to look further into the Universe than our galactic neighbourhood, Cepheid variables are simply too faint to see. Instead, we turn to another standard candle that we have met before, a particular type of exploding star called a supernova 'Type Ia'. These are so luminous they can be observed even in galaxies in the distant Universe.

A Type Ia is a class of supernova that originates from a binary stellar system: two stars in orbit around each other. At the end of the lifetime of one of the stars, the core might collapse into a white dwarf, comprised mainly of carbon and oxygen and swimming with electrons. The white dwarf remains stable because of the Pauli Exclusion Principle, the quantum effect preventing electrons from being in the same 'quantum state'. Basically it means that electrons cannot be squeezed too closely together, resulting in a kind of pressure that opposes further gravitational collapse. At least up to a point.

Left to its own devices the white dwarf would sit there quite happily for an extremely long time. The fly in the ointment is the companion star. This next-door neighbour is a big ball of burning gas, close enough that the white dwarf can gravitationally accrete material from the companion's outer layers. This accretion increases the mass of the white dwarf, in turn increasing the density and temperature of its core. Above a critical threshold, when the white dwarf reaches a mass equivalent to 40 per cent more than that of the Sun, a dramatic event takes place: the temperature of the core allows carbon and oxygen nuclei to suddenly fuse in a nuclear reaction. This releases enough energy to rip through and explosively devastate the white dwarf. This is the supernova.

If you were to write down a number to describe how much energy the supernova releases in watts, it would be a one followed by forty-four zeros. A reasonable fraction of this energy is in the form of electromagnetic radiation: so many photons are released that a single supernova can briefly outshine most of the surrounding galaxy. Over the days and weeks after the supernova, the shattered, expanding remains continue to glow through the radioactive decay of new elements formed in the extreme conditions of the explosion. Like a dying ember, we can see the radiation slowly fade away in a characteristic 'light curve'.

Because all white dwarfs explode at the same critical mass threshold, it is thought that Type Ia supernovae all have the same intrinsic luminosity, because there is a relationship between the amount of mass present and the amount of energy liberated in the thermonuclear explosion. This is why a Type Ia can be used as a standard candle.

In our own galaxy we can expect about one Type Ia supernova to explode every four or five centuries or so. They are reasonably rare events to occur in any one galaxy, at least on a human timescale. Luckily there are hundreds of billions of galaxies in the observable Universe, so to discover new supernovae on a regular basis, one just needs to look at a large number of galaxies. By monitoring them frequently we can search for the sudden appearance of a new point of light that betrays the explosion. There have been many observational campaigns to do just this, and they have discovered large numbers of Type Ia supernovae exploding in galaxies in the distant, high-redshift Universe.

When high-redshift Type Ia supernovae were placed on the Hubble Diagram in the late 1990s, now substantially updated and expanded since Hubble's original graph, something unexpected was seen. The supernovae at high-redshift were fainter than they should be if the local Hubble law of constant expansion held across the history of the Universe. If we trust their validity as standard candles, then they must be further away than would be expected for a constant rate of expansion. This observation was evidence that the rate of expansion must not be constant, or decreasing, but accelerating.

This was the moment that 'dark energy' became an empirical fixture of our cosmological model, for this is the name given to the mechanism driving the acceleration. It is a key component of our standard model of cosmology, but as yet it is beyond our standard model of physics. We don't know what it is.

In our cosmological model, dark energy is indicated by the Greek letter Λ, and this comes from Einstein's field equations. Solutions of the field equations can be expressed in terms of the metric that describes how events in spacetime are separated. In the case of the Universe as a whole, we can find a metric where spatial separations depend on cosmic time, and this can encapsulate the expansion history of the cosmos. This is where general relativity and cosmology meet.

Einstein formulated his field equations prior to 1920, before Hubble had shown that we are living in an expanding Universe. At the time, it was thought that the Universe was static. But the field equations in their original form predicted a Universe in which space was dynamic: in fact contracting due to gravity. So, in keeping with the thinking of the day, Einstein introduced a new term into the equations, given the symbol Λ. This was a mathematical tweak that would oppose gravitational collapse, allowing for a static Universe. This Λ became known as the 'cosmological constant'. Later, when it was observed that the Universe was indeed dynamic, as revealed by redshifted galaxies, Einstein is rumoured to have referred to this modification as his 'biggest blunder'. As it turns out, the cosmological constant *is* needed in the field equations because it allows for a metric solution that describes an accelerating expansion of space.

The Friedmann–Robertson–Walker (FRW) metric, named after the scientists who derived it, was found to be an exact solution to the field equations that also satisfied the fundamental cosmological tenets that the Universe should be homogeneous and isotropic on large scales – that is, its contents should be fairly uniformly spread out and look the same in all directions. Importantly, in this solution the spatial part of the metric can change with time.

This FRW metric is the foundation of our standard cosmological model, and from it we derive the functional forms for the evolution of the scale factor, the redshift and so on. It fully describes the expansion history of the Universe. But while the form of the FRW metric satisfies the field equations; it allows for different evolutionary scenarios that depend on the exact values of parameters – such as the matter density of the universe and the cosmological constant Λ. The objective of observational cosmology is to find the values of the correct parameters. If we know the parameters, we know what sort of Universe we live in and can predict its fate.

Friedmann showed that the metric could be used to describe the expansion of the Universe in terms of a hypothetical perfect fluid that fills it. This fluid can be described by a so-called equation of state, which is a mathematical way of relating the density of the fluid to its pressure. With the discovery of the accelerated expansion through Type Ia supernovae in the 1990s, it was clear that the cosmological constant term Λ is non-zero; it has a particular value. When a positive Λ is considered in Friedmann's equations (as they are now known), something remarkable is revealed: the fluid has a *negative* pressure. It means the accelerated expansion can be considered to be caused by something that contributes to the overall energy density of the Universe, and this contribution has a similar effect to a fluid with negative pressure: it pushes space apart. We don't know what the origin of this additional energy density is, or even exactly what it is, so we call it dark energy.

We are now moving into a phase of the Universe that is 'dark energy-dominated', which means that the total energy density of the Universe is mostly in the form of the mysterious dark energy. This will continue to accelerate the expansion of the Universe forever, as far as we can tell. The observational impact is bleak: light from the distant Universe will get fainter and fainter and redshifted to ever lower frequencies, and the distance between galaxies not already strongly attracted to each other through gravity will forever increase. The lonely cosmos will go dark for each observer.

In such an accelerating Universe, large-scale gravitational potentials – the broad undulations in spacetime – 'decay' over time.

Thinking of the rubber sheet analogy, a large-scale depression in the sheet will get flattened out as if a lot of unseen hands are pulling on all the edges. That's a bit like the effect of dark energy on the large-scale potential wells of the most massive structures in the Universe.

We have already met clusters of galaxies, the largest gravitationally bound objects in the Universe. Clusters are the rare, high-density peaks in the matter density field, but they do not sit at the apex of the hierarchy of structure. Clusters themselves can cluster together, along with their interconnected filaments and neighbouring groups and individual galaxies, into giant structures called superclusters. The most famous supercluster in the nearby Universe is called the Shapley Supercluster, named after the astronomer Harlow Shapley (1885–1972). It contains around ten thousand galaxies and has a mass equivalent to about ten million billion Suns. In turn, there are large volumes of the Universe that are correspondingly rarefied in matter density, nearly empty of galaxies. These are the supervoids.

We can think of superclusters and supervoids as the sweeping valleys and rolling hills of the fabric of spacetime. Like the smoothing of a crumpled sheet, dark energy will tend to flatten them out as cosmic expansion accelerates, because the gravitational collapse of structure on these large scales is slower than the rate of expansion. These structures are so gigantic that the time of flight of a cosmic microwave background photon passing through them is extremely long. So long that, if the gravitational potential is decaying due to cosmic expansion during the crossing time, a photon doesn't need to expend as much energy when it eventually climbs out of the potential as it gained when it first entered. It means that in a Universe dominated by dark energy, we expect to see a typical gravitational blueshifting of the cosmic microwave

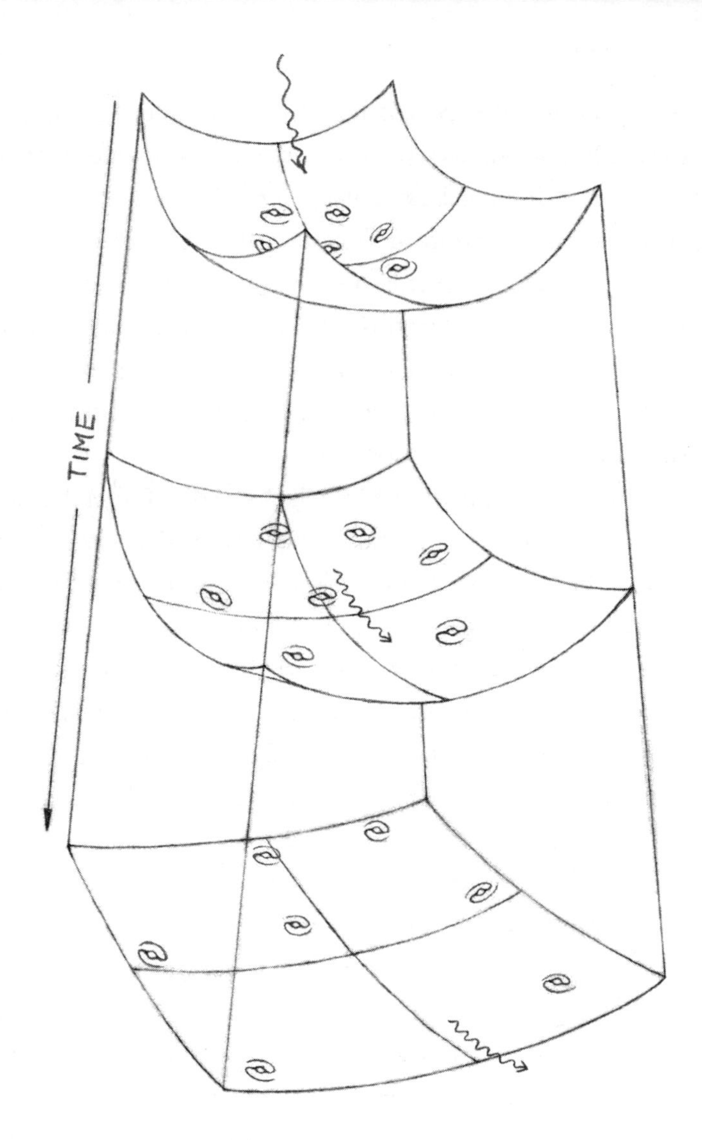

Integrated Sachs-Wolfe Effect

The Integrated Sachs-Wolfe Effect: cosmic microwave background photons travelling through a large-scale structure, like a supercluster, undergo a gravitational blueshift as they enter the gravitational potential, and a gravitational redshift as they leave. Dark energy, driving the accelerated expansion of space, can cause the potentials to flatten out during the time it takes the photon to cross the potential. This means the photon doesn't have to expend as much energy on its outward trajectory out as it gained on its way in – we see a net blueshift.

background towards superclusters. Observationally, the cosmic microwave background looks a little bit hotter as it emerges from a supercluster that has been flattened out by dark energy. The opposite is true of photons travelling through the decaying 'potential hill' of a supervoid: they lose energy on their way into the void, but don't recover this on the way out because the hill has been squashed in the meantime. These photons pick up a net redshift as a result, and the cosmic microwave background looks a little cooler towards the supervoid.

This is the *Late-time* Integrated Sachs–Wolfe Effect, because the impact of dark energy is starting to kick in fairly late into the lifetime of the Universe. Observationally, it is quite hard to convincingly measure the Integrated Sachs–Wolfe Effect directly because it can only be detected over significant fractions of sky. It requires a combination of large-area maps of the cosmic microwave background *and* detailed knowledge of the distribution of galaxies. Without the latter, it is hard to tell where the superclusters and voids are. To make matters worse, these structures are rare – being the most extreme cosmic environments, there simply aren't many of them within our observable Universe.

Despite the challenge in observing it, for me the Integrated Sachs–Wolfe Effect is a more fascinating probe of dark energy than observations of baryonic acoustic oscillations or supernovae. It speaks to the sheer scale of the Universe, both in time and space. On cosmological scales cosmic microwave background photons are leisurely travellers, witness to the slow heave of spacetime as it contorts with the gravitational growth of structure, only to be smeared out as dark energy takes hold, shaping the story of our Universe on its largest scales. Writing its message in the journey of light.

FIVE

BLACK HOLE BEACONS

Nearly everyone who reads this will have heard of black holes. Whenever I give an astronomy talk to a general audience, black holes always come up in the questions. Mention of them gets attention, and rightly so. Their very name suggests some sort of terra incognita. Most of the focus is usually on their intense gravitational field and the idea that matter can fall into a black hole and apparently disappear from the Universe. What is perhaps less appreciated is that black holes can power some of the most intense light sources in the Universe, outshining entire galaxies.

As we have seen, general relativity, expressed through the field equations, describes how matter and energy density deform spacetime. We can think about the strength of the gravitational field around an object in terms of the curvature of spacetime. Not long after Einstein presented his field equations, other scientists started finding 'solutions' to them that describe the behaviour of spacetime under different hypothetical physical situations. One of these scientists was Karl Schwarzschild (1873–1916), who considered the properties of spacetime around an object where all the mass is in a single, ultra-dense point. Schwarzschild showed that in this situation the solution to the field equations becomes infinite, or singular, at a particular radius around the point mass. That special radius has become known as the Schwarzschild radius and within it lies a 'singularity'.

The Schwarzschild radius defines an imaginary spherical shell, like an invisible boundary, around a gravitating body. It has a special property: within the boundary no light can escape back into space. Remember, photons travel along paths in curved spacetime, but within the Schwarzschild radius the curvature is too extreme, like going over the lip of an infinitely deep well with sides too steep to climb. We call that boundary the 'event horizon'. As observers looking towards a singularity, we cannot see past the event horizon, and any source of radiation on the other side cannot reach us. It appears as a 'black hole' – a term coined later in the century for such objects. At first, this was a theoretical exercise: exploring the implications of general relativity for different physical scenarios. But do such objects exist in nature?

To create an object like a black hole requires gravitational contraction to continue unabated, crushing matter into a dense point. Since all mass in the Universe is gravitating, and gravity is 'on' all the time, why don't all massive bodies just collapse into themselves, forming black holes everywhere?

The answer comes from other forces that can oppose gravitational collapse. Stars are perfect examples of this phenomenon in action. Although stars are incredibly massive objects, they can remain stable for billions of years. They do not quickly collapse into black holes, because of the thermal and radiation pressure resulting from hydrogen burning, which act to oppose gravitational collapse. With the 'outward' and 'inward' forces in equilibrium, the star can remain in a stable state – neither collapsing nor exploding – for as long as it can produce enough internal pressure to balance gravitational contraction. When the fuel starts to run out, this happy equilibrium falls apart.

We have already learned how the cores of some stars can collapse into compact objects called white dwarfs at the end of

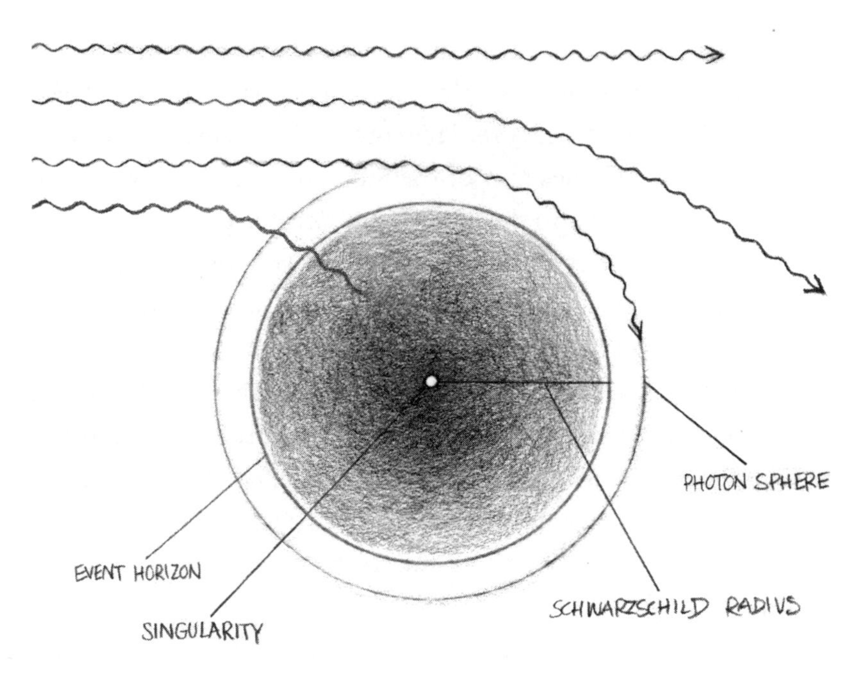

Schwarzschild radius
The Schwarzschild radius describes a boundary around a spherical mass where the solution to Einstein's general relativistic field equations becomes infinite, or singular. This defines the event horizon, within which the curvature of spacetime is too extreme for light to escape. Within the event horizon lies a black hole.

their lives. In these objects the nuclear reactions in the star have ceased and the outer layers of the star have been ejected, leaving a compact, dense stellar remnant made of heavy elements like carbon and oxygen. This dense medium swarms with electrons. The 'degeneracy pressure' between those electrons due to the Pauli Exclusion Principle prevents the white dwarf from collapsing further. This new form of pressure is now opposing the gravitational contraction, keeping the white dwarf stable. Our Sun is destined to become a white dwarf some five billion years from now.

But not all stars retire this way. A star with a mass of about ten to thirty times the mass of the Sun tends to explode as a 'Type

II' supernova at the end of its life, an event that violently ejects most of the mass of the star into the interstellar medium but also leaves behind a dense core. This core can have a mass equivalent to a couple of times that of the Sun, but contained within an object comparable to the diameter of Greater London.

Such an object can overcome the electron degeneracy pressure, but it runs up against something else: *neutron* degeneracy pressure. Most of the matter in this stellar remnant is in the form of neutrons, which, like electrons, also obey the Pauli Exclusion Principle. We call it a neutron star. As before, the degeneracy pressure prevents gravity from contracting the neutrons closer together, supporting it against further collapse. But again, this balance only works up to a point. If the progenitor star is more massive still, then the compact remnant left behind after the star's death can be too massive to be supported even by neutron degeneracy pressure. With no mechanism to prevent the gravitational collapse, the object contracts into a singularity – a black hole. We can see white dwarfs and neutron stars in our galaxy, but what about black holes?

A celebrated example of a real black hole in nature is the object Cygnus X-1. As the name indicates, it is an object in the direction of the constellation Cygnus, and the 'X' signifies that it is a source of X-rays. As we know, X-rays are very high-energy photons, typically associated with extreme astrophysical environments. In this case, the X-rays originate from hot gas being pulled from a blue supergiant star. The star is part of a two-body system where the other member is a compact object with a mass of about fifteen times that of the Sun. These objects orbit each other every six days, and as a whole the system is known as an 'X-ray binary'. It is thought that the compact object is the remnant of another massive star which, at the end of its life, might have collapsed directly into a black hole. Of course, by their very nature, black holes are rather hard to detect, but Cygnus X-1 betrays its presence through its gravitational effect on its companion star, which we *can* measure, and the flicker of X-rays emitted by the hot gas between them.

Over many years of observations, we now have a robust list of black hole candidates, and there has recently been an exciting breakthrough in this field: the ability to detect *gravitational* waves emitted by the merging of two black holes. Gravitational waves are distortions – ripples – in spacetime that propagate through the Universe; another prediction of general relativity. When two black holes (or other compact massive objects, like neutron stars, for that matter) merge, they undergo a spiral dance, orbiting each other faster and faster until they coalesce to form a single black hole. This dance and final crescendo cause a pattern of ripples in spacetime to propagate outwards into the Universe. The new, merged black hole is massive, but not quite the sum of the mass of the two individuals; some of the total mass of the system is radiated away as gravitational waves.

When these ripples pass through the Earth, they cause tiny distortions in our local spacetime that we can actually detect as minute changes in the length of well-calibrated measuring rods. Thus we can now build gravitational wave telescopes. This new realm of gravitational wave astronomy is fantastically exciting because it is like opening up a brand-new sense with which to understand the Universe, and now there are observations of several black hole mergers on record. Such is the importance of this breakthrough that the direct detection of gravitational waves won the Nobel Prize in Physics in 2017.

So despite how difficult they are to detect, we are very confident that black holes are real objects in the Universe. In fact, black holes are rather common, and it turns out they play a key role in the story of galaxy formation, for in the 1990s it was discovered that every massive galaxy contains a big black hole at its centre. Our own galaxy is no exception.

The centre of the Milky Way lies in the direction of the constellation of Sagittarius, visible from the southern hemisphere. Our galaxy is shaped like a disc; at its hub is a central bulge of stars that is like the yolk of a fried egg. Buried in the heart of this 'yolk' lies our own central galactic black hole. Such an object was predicted to exist when astronomers first considered how galaxies might have formed and evolved from primordial conditions, but how do we know it is there?

Taking images of this part of the galaxy in the visible bands of the electromagnetic spectrum is problematic because we are peering into its densest port, crowded with stars and thick with interstellar gas and dust. One way of looking through this clutter is to use slightly longer wavelengths of light. If we use a camera sensitive to near-infrared electromagnetic radiation, with wavelengths of one or two microns, we can see through the mess. The slightly lower energy near-infrared photons pass more easily through the dust clouds blocking our view because they have a lower probability of being absorbed or scattered by interstellar dust than those photons at visible or shorter wavelengths.

Still, we also have to contend with the stars themselves. The central part of our galaxy is extremely cluttered with them, and so we have to peer between stars too, like looking towards the centre of a forest through the trees. To help, we can make images sharper using a technique we met before, called adaptive optics, which compensates for the blurring effect of the Earth's atmosphere. By monitoring a bright reference star, or a bright spot in the sky made by a powerful laser, it is possible to rapidly adjust the shape of a telescope's mirrors in such a way that the distortions in the paths of incoming photons caused by our atmosphere are countered. Two closely separated stars that appear blended together in an image because of this blurring suddenly become crisp and distinct. We can see through the trees to the centre of the forest.

So when we manage to take a sharp image of the very centre of our galaxy, can we see any evidence of the black hole? At first glance, no, at least not in visible and near-infrared light. But wait a few years and look again, and then again and again, and a remarkable picture emerges. Some of the stars at the centre of the galaxy are apparently moving on orbital paths: ellipses focused on a common unseen central point. This unseen point is the location of a black hole. It is invisible but betrays its presence though a gravitational influence on the stars around it. We call it Sagittarius A* (pronounced 'A-star'). The stars' orbits tell us how much mass is present in that unseen spot: about four million times the mass of the Sun. We class it as 'supermassive'.

We can indirectly detect, and weigh, supermassive black holes in other galaxies by looking at the motions of the stars and gas at their centres. Although we cannot track the motions of individual stars over time like we can in the Milky Way, we can measure the combined *distribution* of the speeds of stars or gas in other galaxies.

When we observe other galaxies, we typically measure the light emitted by large numbers of stars all at once. If the stars are moving relative to each other within the galaxy, spectral features such as absorption lines (caused by the presence of heavy elements in stellar atmospheres that can absorb a photon with a particular energy) in individual stars will be red- or blueshifted by the Doppler effect. The amount each line is shifted depends on each star's velocity, relative to the average: some will be moving towards us, some will be moving away, and there will be a distribution – a particular spread – of speeds.

When we disperse the light of a distant galaxy into a spectrum, we see the same set of absorption lines expected for individual stars, but all the individual stellar lines that have been Doppler shifted by different amounts around the average get blended together. This broadens the widths of the 'total' absorption lines we observe in the total galaxy spectrum. In turn, these widths, measured in frequency or wavelength, can be translated to the spread in the distribution of velocities of the stars in that galaxy. We can do this for whole galaxies, or for parts of galaxies, so we can 'map' the motions of stars, even in distant galaxies.

The presence of a massive, highly compact object like a supermassive black hole causes a particular broadening of the velocity distribution of the stars and gas at the centre of a galaxy, such that the stuff near the black hole is moving very rapidly in a way that cannot be explained through other internal galactic motions. This not only provides us with an indication that a black hole is present, but gives us a handle on its mass, because the velocity dispersion gets larger with increasing central mass. That's one method to weigh black holes in other galaxies.

In the Milky Way, our supermassive black hole sits there like a spider at the centre of a web, not doing much save for patiently exerting its gravitational influence on the surrounding stars and the occasional passing gas cloud. We call it 'quiescent'. But imagine a situation where the black hole was continuously 'fed' with matter: what would happen then? A feeding black hole can actually turn into a cosmic beacon, emitting so much energy that it can outshine the entire galaxy around it, emitting light that is visible over nearly the entire observable Universe.

In 1963 the astronomer Maarten Schmidt, working at the Palomar Observatory in California, observed an object called 3C 273, a strong emitter of radio waves. Visible light images revealed the source to be like a star: a bright, unresolved point of light. If it *was* a star, it should not be such a strong radio source, and it should also be located within our galaxy. But the spectrum of 3C 273 revealed that the key emission features were at much longer wavelengths than they should be if the source was local. In other words, the spectrum showed a strong redshift, indicating that the light Schmidt was detecting was extremely distant, with a light travel time of two billion years. This made 3C 273 the most distant source known at the time and because of its brightness implied that the object emits remarkable amounts of energy.

Objects like 3C 273 became known as quasi-stellar radio sources, or 'quasars'. Even today, quasars are our best probe of the deep cosmos, shining across the Universe like beacons among the comparatively faint galaxies.

Quasars are a type of galaxy in which the central supermassive black hole is actively feeding, or as we say, 'accreting'. Like a parasite burrowed deep within, it is pulling in interstellar gas from its host and increasing in mass. You would be forgiven for thinking that when matter crosses the event horizon and falls into a black hole it simply vanishes, lost to the Universe. But it's not as simple as that. It is the process of accretion – what happens to the matter *as* it falls in – that releases huge amounts of energy and causes the black hole, or rather its immediate environment, to shine.

If I stand on a diving board high above the water, my body has what we call gravitational 'potential' energy. The amount of potential energy is given by the product of my mass, the acceleration due to Earth's gravity and my height above the ground.

This is where the 'potential' in potential wells comes from. There is a gravitational force pulling me down towards the centre of the Earth, but while my feet are on the diving board there is an equal and opposite force pushing me up, keeping me in place. The moment I step off the diving board, that upward force vanishes and the gravitational force starts accelerating me towards the water.

Close to the surface of the Earth the acceleration of any object due to gravity is an increase in velocity of about 10 metres per second every second. So after one second I am falling at a velocity of 10 metres per second, after two seconds I am falling at 20 metres per second, and so on. My potential energy is dropping as I approach the water, because my height above the surface is decreasing. But my kinetic energy is increasing – I am hurtling through the air faster and faster as I fall. So we can think of this as a transfer of potential energy to kinetic energy. It is this idea of the conversion of energy and its transfer from one form to another, and how it affects objects in the Universe, that underpins a great deal of physics. A fundamental tenet is that energy cannot be made or destroyed, just converted between forms.

It follows that when I hit the water and slow to a stop, I lose my kinetic energy. Where does it go? In this case that kinetic energy is dissipated in the water: I transfer energy to the water molecules, violently pushing them around and creating a splash. Each drop will carry away a small bit of my kinetic energy, which in turn gets dissipated further as the splash settles. Waves emanate from where I enter the water, carrying energy away. That energy has not gone from the Universe but has instead been diluted over the water molecules in the pool.

Any clump of material within the vicinity of a supermassive black hole has gravitational potential energy. If the material is to be accreted onto the black hole, this potential energy must be lost – converted to another form.

The black hole can gravitationally attract the matter in its vicinity, but this does not tend to flow towards the black hole on a direct radial path. Generally, the local interstellar medium will have some bulk rotation – in a disc galaxy like our own for example, the gas in the disc is rotating around the central hub. In some galaxies, the central black hole might be surrounded by an extensive, dense reservoir of orbiting gas.

When we think of the dynamics of an orbiting body, angular momentum is a fundamental property. It is defined as the mass of the body (say, how many atoms are in a cloud of gas) multiplied by the radius of its orbit and by its speed. Another fundamental physical concept is the *conservation* of angular momentum. If a gas cloud is orbiting a black hole with some angular velocity, and we reduce the radius of its orbit, the circular speed must increase in order to conserve angular momentum. The same principle is why a spinning ice skater will spin faster and faster as they draw in their arms: their angular momentum must be conserved.

If there is lots of gas around a supermassive black hole with bulk angular momentum, the orbiting material can be drawn closer. As it does so the conservation of angular momentum makes the gas orbit faster and faster, forming an 'accretion disc', like a spinning pizza dough. It is from this accretion disc that our photon emerges, and the point supermassive black holes start shining.

Frictional and viscous forces in the dense accretion disc cause the gas to heat up. It heats up so much that the disc starts radiating high-energy electromagnetic radiation, and it glows brightly with visible, ultraviolet and X-ray light – very high-frequency photons. The disc is 'radiating' its energy away. Through these viscous forces, the angular momentum of the gas closest to the black hole is transferred up through the disc, and this allows material from the inner part of the accretion disc to cross over the event horizon. Then it is truly accreted onto the black hole. So this process of accretion allows the central black hole to increase in mass, but releases energy in the process.

The luminosity of the accretion disc – the amount of energy it radiates every second – can exceed all of the starlight from all the billions of stars in the surrounding galaxy. And since the size of the accretion disc is tiny compared to the scale of the galaxy, like a marble compared to the Earth, to an outside observer viewing the galaxy from afar – like us – it is as if all of that energy is emanating from a single point. So when we see a galaxy in which the central black hole is voraciously feeding in this way, they can shine like single point sources, glaring so brightly that we often can't make out the surrounding 'host' galaxy at all.

Being so bright, and therefore easy to detect over vast cosmic distances, it is fairly straightforward to measure the number of quasars in the Universe and track how their abundance has changed with time. Through dedicated quasar surveys, we've known for some time that the peak era for quasar activity occurred about eight to ten billion years ago. This coincides with the peak era for star formation activity in galaxies and is a clue that galaxy growth through star formation, and the growth of their central black holes, are linked.

Clearly the growth of stars and the growth of the central black holes in galaxies are linked in the sense that they both rely on a similar basic recipe: more fuel (interstellar gas) generally means more activity, and there was generally more gas in galaxies in the early universe. But in the 1990s a more nuanced picture

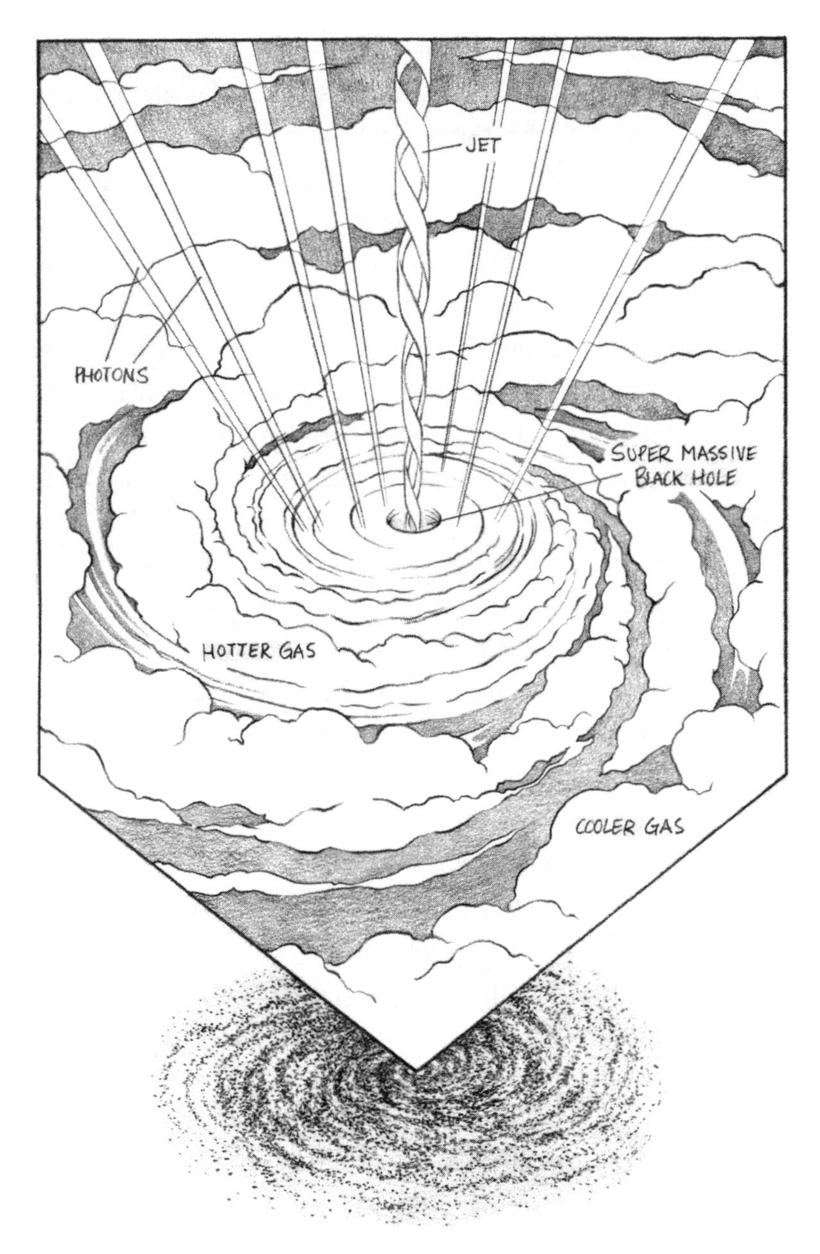

Supermassive black hole accretion disc

A fast particle jet and luminous accretion disc around a central supermassive black hole, formed as the black hole 'feeds' on interstellar material.

emerged. When astronomers started comparing the mass of the central black hole to the mass of the stars in the surrounding galaxy, it was found that the two are quite tightly correlated. At first, this seems pretty intuitive – the bigger galaxies have bigger black holes, right? But is this so obvious? Not when you consider the relative size of the black hole and the galaxy around it. The size of the black hole, or rather, its surrounding luminous 'engine', is many orders of magnitude smaller than the size of the galaxy. That means the mass of the stars and the mass of the black hole should not necessarily be so strongly correlated, unless they were somehow connected by a physical process. It made astrophysicists wonder: how could a comparatively tiny black hole affect the fate of the entire galaxy around it?

The answer lies in the intense energy output of those quasars. The accretion disc doesn't just glow like a light bulb, it can drive a powerful 'wind' from its surface. One mechanism for this to happen is through radiation pressure – the effect of photons slamming into atoms or molecules and transferring momentum. We have met this concept before when thinking about the cores of stars, held up against gravity, and the production of the baryonic acoustic oscillations in the early Universe. Although photons have no mass, relativity tells us that they do have momentum, which is proportional to the energy of the radiation. This momentum can be transferred to particles of matter that happen to be in the path of a photon, such as ambient gas and dust around the centre of the galaxy. Like blowing away a cloud of smoke, this momentum transfer can drive an outflow of interstellar material away from the accretion disc, into and through the wider galaxy around it.

This process of a quasar (or indeed a supernova or even an individual star) transferring energy or momentum into the surrounding interstellar medium is called 'feedback'. Feedback can have two effects on a galaxy. Positive feedback can cause the formation of new stars by triggering clumps of gas to gravitationally contract and collapse. One way for this to happen is if the wind emanating from an accretion disc slams into a cloud of molecular hydrogen, dissipating energy in heat and turbulence in the gas. The injection of turbulence causes ripples of density fluctuations in the cloud, some of which can be large enough to start to fragment and collapse under gravity, igniting bursts of star formation. Left alone, that cloud might have remained quiescent.

Positive feedback can impel gas clouds to collapse and form stars, but feedback can also stop stars from forming. This is called negative feedback. A quasar wind can sweep up interstellar gas as it blows through the galaxy, clearing its path like a snow plough. In extreme cases, that gas can be propelled right out of the galaxy and into the space around it, into an environment called the circumgalactic medium. So, by removing gas that would otherwise be converted into stars deep within a galaxy, feedback from a growing central black hole can dramatically affect the evolution of a galaxy as a whole.

Sometimes star formation might be totally quenched in the short term, but it may well be that the ejected gas may later 'rain' back down onto different parts of the galaxy through gravity, like a fountain, to form new stars in the future. This churning and redistribution of gas within and around galaxies plays a vital role in the chemical evolution of galaxies. In some ways we owe our existence to this gas recycling because it allows heavy elements like carbon and oxygen to be well mixed throughout the interstellar medium. By agitating and mixing interstellar gas, feedback ensures elements produced in stars and supernovae in one part of the galaxy can be spread far and wide, like seeds carried on the breeze. Much later, some of those elements can go on to be part of new solar systems, like our own.

Quasar feedback provides a mechanism for the comparatively minute black hole to affect star formation across an entire galaxy, owing to the colossal amount of energy liberated from the gravitational potential energy of surrounding gas clouds. Of course, by removing gas from its immediate vicinity, aggressive quasar feedback also affects the growth of the central black hole itself: when there is no more gas to feed from, the accretion disc will quickly disappear and the quasar will turn off. If gas later returns to the vicinity of the black hole through gravitational in-fall, or it is brought there through the dynamical motion of the galaxy, or a collision with another galaxy, the quasar might switch back on.

Through this cycle of accretion and feedback, the black hole and galaxy around it are locked in a state of self-regulation, and this concept of regulatory feedback is now a central feature of our picture of how galaxies evolve. Without it, models and simulations of galaxy formation cannot reproduce the correct distribution of galaxies we see around us today.

The light from quasars obviously has a dramatic effect on the innards of the galaxy that hosts it. Inevitably, many of the photons emitted from the accretion disc must be scattered or absorbed internally, never to escape the galaxy. Despite this, many quasars shine very brightly, meaning plenty of photons do escape the galaxy, travelling into the Universe. After they have escaped their galaxy, some of those photons have a long journey to our telescopes.

The space between us and any distant quasar is not empty; it is filled with other galaxies and intergalactic gas clouds. And the further away a quasar is, the more of this stuff there is for the photons to contend with. We can turn this to our advantage, using the quasars to reveal otherwise unseen cosmic matter.

Like all sources of radiation, the quasar accretion disc produces a characteristic spectrum of electromagnetic energy. One of the main features of quasar spectra is a strong 'continuum' of radiation, which means that photons are emitted by the accretion disc across a wide, continuous range of energies, from the X-rays right through to the visible light bands and beyond. These photons of many different energies leave the accretion disc, and those that escape the galaxy set off on a journey through a Universe that is laced with (mostly) hydrogen gas, both in and around galaxies.

An atom of hydrogen can absorb a photon of a very specific energy, corresponding to a wavelength of almost exactly 121.6 nanometres. Quantum mechanics has shown us that electrons reside in discrete energy 'levels' in atoms. It's a bit like the electrons occupy different floors in a block of flats. If you go up a floor, you go up an energy level. The energy of a photon with a wavelength 121.6 nanometres is equivalent to the difference in energy between the ground floor (the 'ground state') and the first floor. An incoming photon of exactly this wavelength can be absorbed by the atom by promoting an electron in the ground state to the next level. Later, that promoted electron can return back down to the ground state, but to do so it must release a photon of exactly the same energy, and therefore the same wavelength, 121.6 nanometres.

Any electron transition to or from the ground state of hydrogen is part of the 'Lyman series', named after the physicist Theodore Lyman (1833–1897). At the start of the twentieth century Lyman discovered the various telltale emission lines of light in the spectrum of energized hydrogen gas. Each has a very specific colour, corresponding to the precise difference in energy of various electron transitions. The transition involving a photon of 121.6 nanometres is called Lyman-alpha, the first in the Lyman series.

If a quasar happens to lie on the same line of sight as a cloud of neutral hydrogen gas in the intergalactic medium, some of the photons from the quasar with energies corresponding to the Lyman-alpha transition might be absorbed by some of the atoms in that cloud. When we disperse the quasar light into a spectrum, we see that the broad continuum of light is punctured with a hole: an absorption line where the photons were stolen by the intervening cloud. And we can do more than simply say that there is a cloud in the way; the size of the dip can be used to estimate the number of hydrogen atoms in the cloud, and therefore we can also determine the cloud's mass.

Both the light from the distant quasar and the intervening cloud of gas are the subject of redshift due to cosmic expansion. So we don't actually observe the absorption line at 121.6 nanometres here on Earth, we see it shifted to a longer wavelength in the quasar spectrum, according to the redshift of the intervening gas cloud. And there might not be just one cloud between us and a quasar, there may be many, all at different distances and therefore at different redshifts. This can riddle a quasar spectrum with Lyman-alpha absorption lines. Each line corresponds to a different cloud of neutral hydrogen between us and the quasar, perhaps widely spaced along the line of sight, like beads on a string. The more distant the quasar, the more of these absorption lines we see. We call this the 'Lyman-alpha forest', because it leaves a forest of absorption lines on quasar spectra.

Quasars give us unique information on the distribution of matter along narrow columns through the cosmic volume. They reveal matter – the intergalactic hydrogen gas clouds – that would

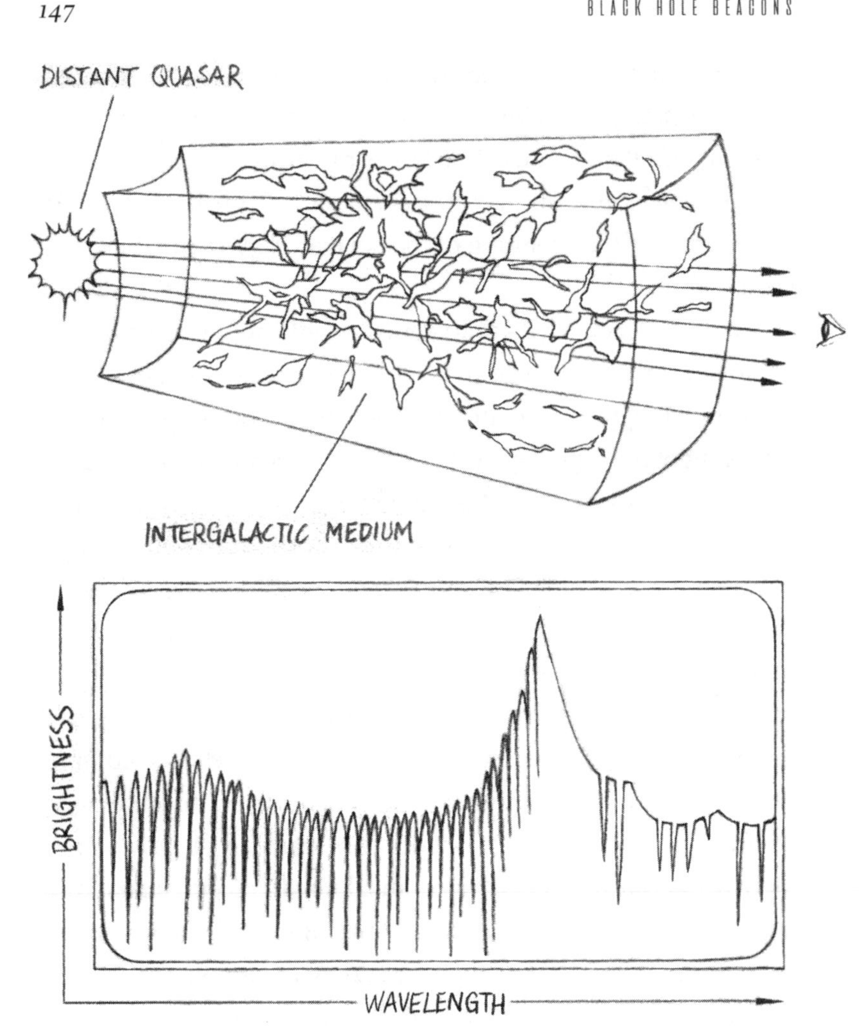

Lyman-alpha forest
The imprint of the Lyman-alpha forest on quasar light, caused by intervening clouds of neutral hydrogen gas in the intergalactic medium.

otherwise be very difficult to detect, because the clouds might not contain many, if any, stars or other luminous matter. Instead, we see them in a kind of silhouette against the bright background quasar. The Lyman-alpha forest is denser at high-redshifts, so we see more absorption lines in the spectra of very distant quasars.

Over cosmic time the forest thins out because a lot of the intergalactic gas is eventually accreted onto other galaxies or forms stars itself. This means that 'absorption line studies' of the Lyman-alpha forest through quasar surveys is a great way to build up a picture of the evolution of the normal, baryonic matter over cosmic time.

As a whole, we have a pretty good handle on the contribution of normal matter to the overall mass–energy budget of the Universe. It's quite a small contribution, accounting for about 5 per cent of the total. The rest is thought to be in the form of dark matter and dark energy. Mass and energy are counted together in this budget because in relativity there is an equivalence between the two, and in Einstein's field equations 'mass density' and 'energy density' both have an effect on the curvature of spacetime.

Of the 5 per cent of normal stuff – atoms and suchlike – we can think about how that might be divided up into different components (like stars, dust and gas) and how these relative fractions have evolved over time. We can of course directly observe things like stars because they emit light, and, knowing something of the physics of stars, we can turn a measurement of the light into an estimate of stellar mass. When we add up the amount of matter in stars and interstellar gas and dust in galaxies in the early Universe and put this together with the amount of intergalactic gas revealed by the Lyman-alpha forest, we can account for the majority of normal matter that was present at the Epoch of Recombination. But as time goes on something strange happens: when we look at the amount of normal matter later on in cosmic history – in other words, the stuff we can see nearby, whose light hasn't taken too long to reach us – we come up short. We start missing normal baryonic matter.

We can't account for something like a third of all the normal matter in the local Universe that we think should be there, based on the primordial baryon budget. It means that a rather significant fraction of all the normal matter formed in the Big Bang is now not in galaxies at all, but in an elusive medium that evades detection. Our best guess is that it is in a fairly hot plasma that surrounds galaxies, snaking through the cosmic web of large-scale structure. We call it the warm–hot intergalactic medium.

Numerical simulations of cosmic structure formation suggest that most of the gas in the intergalactic medium of the early Universe never made it into galaxies at all. Instead, it was accreted onto the vast filamentary dark matter structures that grew out of the primordial density fluctuations. This gravitational accretion caused the gas to gain energy, heating up to hundreds of thousands of degrees in the process, becoming too energetic to be efficiently accreted onto galaxies. Now, this is obviously quite hot by everyday standards, but unfortunately it's not hot enough to emit copious amounts of electromagnetic radiation that we can actually detect with current instruments. It's an electromagnetic blind spot.

If the gas was a bit hotter, say by *tens of millions* of degrees, the electrons zipping about in the plasma would start emitting lots of X-rays, which we can detect with X-ray telescopes such as the Chandra X-ray Observatory. In some places we can detect the hot intergalactic medium through X-rays, but only in very dense environments like clusters of galaxies. The warm–hot intergalactic medium, linking cluster to cluster, and surrounding galaxies, is comparatively tepid, and so it is unseen.

So how do we know this stuff is there at all? Quasars to the rescue. Just as clouds of neutral hydrogen gas can leave an absorption line imprint in background quasar spectra, elements within the warm–hot intergalactic medium can do the same thing. The gas is hot enough that all the atoms within it are highly energized. In fact, they are all heavily ionized, meaning they have had lots of electrons stripped away from them. And we are not just talking about hydrogen. The gas is not pristine; after billions of years it has been polluted with heavy elements formed during star formation and ejected from the galaxies embedded within the filamentary intergalactic medium through the same feedback processes we talked about earlier. These pollutant elements can be used as tracers of the warm gas.

One of these elements is oxygen. A normal, neutral oxygen atom has eight electrons around the nucleus, arranged in their different energy levels. It takes a little bit of energy to strip the outermost electron, a bit more energy to remove the next one, some more still to remove the next one, and so on. Basically, the closer the electron is to the nucleus, the stronger the attraction from the protons and the harder it is to remove.

In the warm–hot intergalactic medium, the energies are high enough for most of an oxygen atom's electrons to be stripped away. The resulting oxygen ion can still absorb photons, but this happens at a correspondingly higher energy, meaning that the ionized oxygen absorption lines appear at much higher frequencies in the quasar spectrum than the Lyman-alpha absorption lines. Absorption lines associated with ionized elements in the warm–hot intergalactic medium are typically deep in the high-energy ultraviolet and even X-ray part of the spectrum. Luckily, some quasars can be so bright even at these high frequencies that the absorption lines can still be detected.

But still, we only have glimpses. It's like trying to paint a picture of a rich landscape, but from within a dark box with a few holes punched here and there that let the daylight in. We can only see the gas along the lines of sight towards those relatively few bright background quasars that happen to have a suitable clump of enriched warm–hot gas in front of them. There are simply not many of these fortuitous lines of sight. Although we have mastered nearly every form of electromagnetic radiation, continually devising technology to capture, detect and record it, the warm–hot intergalactic medium remains a shadowy part of our material Universe.

That this intergalactic medium contains a substantial fraction of the normal matter in the Universe – the same basic stuff that you and I are made from – should give us pause for thought. We humans, and indeed the sum of all sentient life that has ever existed anywhere in the cosmos, are simultaneously a vanishingly insignificant component of the Universe and perhaps also its most complex product. It reminds us that despite how far we have come in our understanding of the Universe we inhabit, there is still much to learn. Much we simply cannot yet see.

SIX

RADIO FROM THE COSMIC DAWN

We are at a frontier. One of the ultimate goals of astrophysics is to detect the light from the cosmic dawn, when the first stars burned in the first galaxies. We're not there yet. The light from the most distant galaxy we have detected to date has taken over thirteen billion years to reach us: we see it as it was just a few hundred million years after the Big Bang. It is among the first galaxies for sure, but as we look deeper, we are certain to find more galaxies even further back into the Universe's history. They are just very hard to spot. Actually, what we are really interested in is when do we *stop* detecting galaxies?

Why do we care about this? Observationally, the 'cosmic dawn' is the missing chapter of the story of galaxy formation and represents a significant shortcoming in our empirical understanding of the Universe.

We have a pretty good idea of how the Universe has evolved and what it is made of, even if we don't fully understand its main components, dark matter and dark energy. We know what sort of conditions prevailed before galaxies formed, even if we don't know how the Universe came into existence. We have pinned down the broad evolution of galaxies over the majority of cosmic time, even if there are many details still to learn about their astrophysics. But the time between the formation of atoms of hydrogen at the Epoch of Recombination and the ignition of the first stars is truly terra incognita. It is a Dark Age.

What we do know is this: an important transition occurred as the first stars ignited and quasars turned on. The dark, electrically neutral sea of hydrogen gas formed at recombination was illuminated from within by the electromagnetic radiation shining out of the first galaxies. Of course, the hydrogen was responsible for the formation of these sources of light: it flowed into clumps of matter that were spread throughout the cosmos, the seeds that were set down in a characteristic large-scale pattern, originally imprinted as quantum-level density perturbations at the very start of the Universe.

Ultraviolet photons streamed out of the first stars. Some of the photons, those with an energy in excess of the binding energy of the electrons in hydrogen, collided with the surrounding atoms, removing electrons. The neutral hydrogen atoms were ionized. Light spread like a disease. Around each new star spread a bubble of ionized gas, growing like an out-of-control cell culture as the photons raced away from their origins into the dark Universe. Eventually, nearly all of the neutral hydrogen atoms in the intergalactic medium were ionized. We call that era the Epoch of Reionization, because the normal matter in the Universe returned to a nearly completely ionized state, as it was before the Epoch of Recombination.

Without even detecting the stars and galaxies responsible, we have reasonable observational constraints on when this transition happened. By definition, reionization produces a population of free electrons that are liberated from the hydrogen atoms. The average cosmic density – that is, the typical number of particles per cubic metre of space – of these electrons at the start of the Epoch of Reionization is determined by the production rate of free electrons and the size of the Universe at the point the first stars switched on. Eventually the average electron density decreases

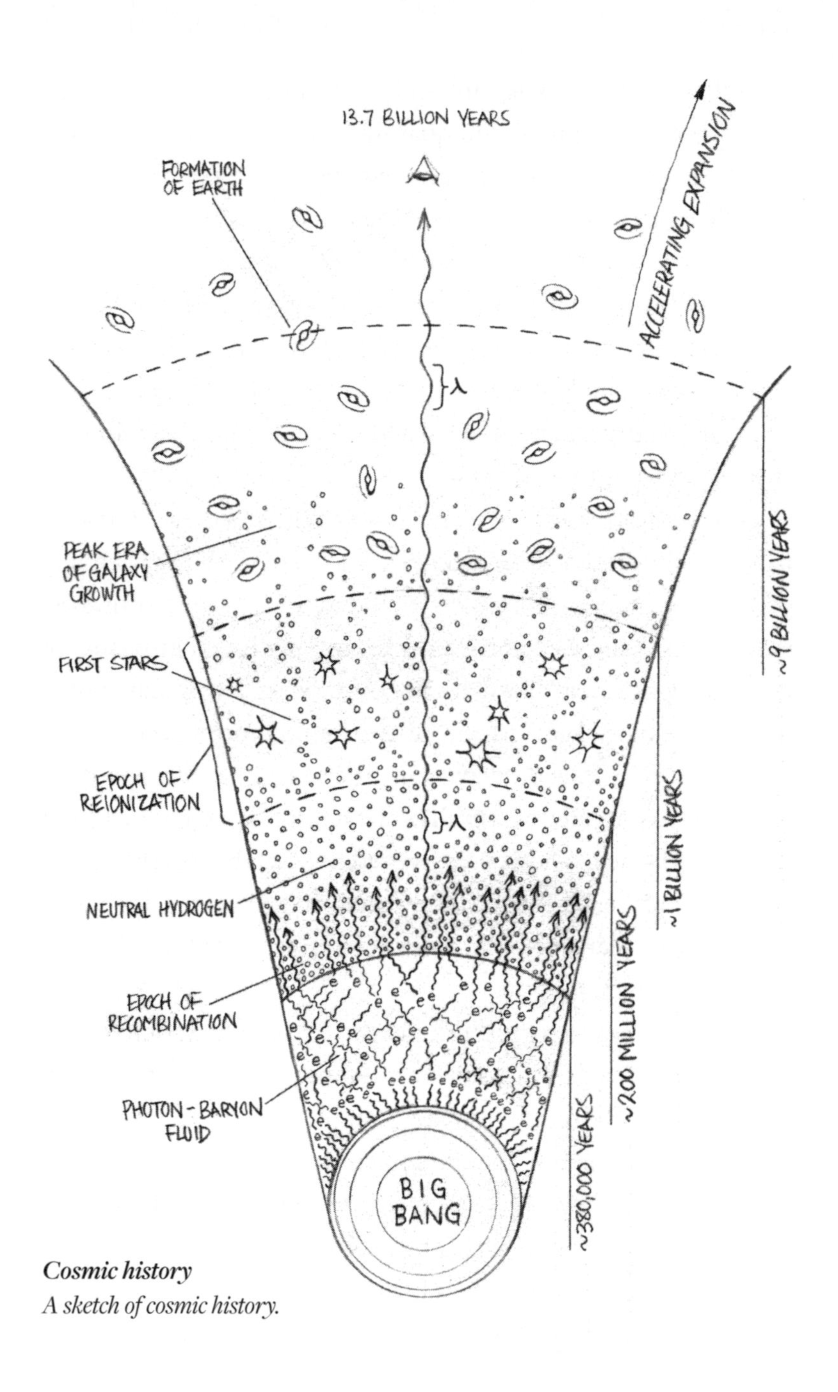

Cosmic history

A sketch of cosmic history.

over time as all the electrons are freed and the Universe continues to expand.

The appearance of free electrons during reionization means that space isn't completely transparent to cosmic microwave background photons; some of them interact with the fresh population of free electrons. As was the case prior to the Epoch of Recombination, those free electrons scattered some of the cosmic microwave background photons travelling through them. This introduces a noticeable effect, akin to looking at the 'true' microwave background through a slightly tinted window. Observationally, this slightly dampens the amplitude of the cosmic microwave background temperature anisotropy power spectrum – the distribution of the hot and cold fluctuations in the background – compared to that expected from a clear, electron-free, line of sight. The line of sight is slightly opaque.

Observations of the cosmic microwave background by the Wilkinson Microwave Anisotropy Probe and Planck satellites have constrained this opacity, allowing us to estimate when reionization commenced. Current data suggests it started no later than a few hundred million years after the Big Bang, comparable to the most distant galaxies we can currently see, but it is likely that there were nascent stars and galaxies contributing to reionization even earlier than this, perhaps within just one hundred million years of the Big Bang.

Observations of large samples of galaxies and quasars seen at later times – that is, with small redshifts – show that reionization must have completed by about one billion years after the Big Bang. We can tell this because, if there were large amounts of neutral hydrogen still present in the intergalactic medium, then the ultraviolet photons emitted by galaxies and quasars would be readily absorbed, as we see in the thickening Lyman-alpha forest.

After the completion of reionization, the Universe became nearly totally transparent to most forms of electromagnetic radiation.

So you can imagine the Epoch of Reionization as a time span of about a billion years during which the Universe gradually burst into light as the first stars in the first galaxies were kindled, like buds blossoming with the flush of Spring. The first generation of stars are thought to be rather different to the stars we see in all the galaxies we have so far observed. Most strikingly, these first stars were born far more massive than most stars today, with masses up to several hundred times the mass of the Sun. They had correspondingly high luminosities. Confusingly, these are referred to as 'Population III' stars. It's another unfortunate bit of astronomy nomenclature, I'm afraid. I'll try to explain.

When astronomers started looking at the stars in our galaxy as well as in other galaxies, it was noted that they had different properties and could be distinguished into distinct stellar 'populations'. Moreover, these different populations seemed to inhabit particular parts of the galactic disc.

Besides a star's colour and luminosity, one of the most important properties is the 'metal' content, or metallicity, of a star. To an astrophysicist, any element heavier (that is, containing more protons and neutrons) than hydrogen and helium is referred to as a metal. It is really just a lazy way of referring to elements that were not formed during nucleosynthesis shortly after the Big Bang. Most of the other elements are formed *in* stars during stellar evolution, or in the explosive ejecta of supernovae, or other violent events. Gold, for example, can be formed through the merging of two neutron stars.

When we look at the spectra of stars we can usually see narrow, dark lines in the rainbow continuum of light. These correspond to the absorption of photons by elements in the stellar atmosphere. Different elements have different numbers of protons, neutrons and electrons, and each element can absorb photons with a range of very specific discrete energies, corresponding to the different quantum energy transitions within the atom. Add all the absorption lines from all the different elements together and you get a kind of barcode of absorption lines in the stellar spectrum that tells you which elements are present – carbon, magnesium, calcium and so on. It gives us another way of classifying stars.

It was quickly noticed that stars in the central bulge of our galaxy had fewer metals in them, and were typically cooler, than those in the main disc of the Milky Way, like our Sun. This means that the bulge stars are, on average, older than the disc stars. We know this because the massive, hotter stars in the bulge population are short-lived and therefore died off long ago, leaving behind the lower-mass, cooler and long-living stars. It also means that these bulge stars generally formed earlier in cosmic history, born in clouds of hydrogen gas that contained comparatively few metals. Metals can only be formed during stellar evolution, or from stellar explosions or the collision of stellar remnants. They take time to build up in the interstellar medium, so there are simply fewer metals as you look back into cosmic history.

By contrast, new generations of stars born in clouds of gas that have been 'polluted' (a politer term we use is 'enriched') with elements tend to have more metals in them. These elements come from previous generations of stars, dispersed through galaxies as part of the regular churn of interstellar gas. The Sun is one of these stars: our Solar System formed from a cloud of hydrogen gas that was enriched by previous generations of stars and supernovae. The Sun contains some of those elements, and some of them have coagulated into planets and people.

The younger, metal-rich stars like our Sun are called Population I stars. The older, metal-poor stars in the bulge are called Population II stars. The very first generation of stars formed from gas that had not yet been enriched with heavy elements – in other words, cosmologically 'pristine' gas. Therefore the first stars should contain no metals. To fit into the nomenclature, they are called the Population III stars.

A dearth of metals in the primordial gas clouds allowed Population III stars to be far more massive than Population I and II stars today. As we know, to form stars, a cloud of gas must 'cool' so that it can fragment into self-gravitating clumps; these go on to gravitationally collapse and ignite nuclear reactions. Cooling means that thermal energy is removed from the atoms or molecules in the gas cloud, and one way for this to happen is for those particles to radiate photons. For example, atoms can release photons when their electrons make transitions between excited quantum states, and this can be triggered by collisions between atoms within the cloud. The photons carry away the thermal energy into space, and the cloud cools.

For a given energy transition in a particular element, this radiative cooling tends to happen only for certain environmental conditions, determined largely by the local density and temperature of the surrounding gas. But when different elements are mixed in with the hydrogen, there are far more possibilities for the gas cloud to radiate energy. There are simply more ways for the atoms to release photons and cool the cloud.

Imagine this cooling like trying to clear a very tall bookshelf. On your own you can remove the books within your reach, but you have to stoop and stretch to remove the books on the bottom shelves, and you can't reach the ones way up high. But with a group of friends, some tall, some short, you can work together to quickly clear all the shelves. In other words, a gas cloud can cool far more efficiently when it is polluted with metals. In metal-free gas, as was present in the very early Universe, cooling cannot happen rapidly. As a result, the cloud tends to fragment into larger clumps, which go on to self-gravitate into very massive individual stars.

Being so massive, these stars had short lives, quickly burning through their gas. Some will have exploded as supernovae at the end of their short lives, dispersing any new elements they forged. Some of these stars may have collapsed directly into black holes at the end of their lives. In fact, this is one of the theories of the origin of supermassive black holes in the centres of galaxies. If the very first galaxies contained a number of moderately massive black holes, spawned from Population III stars, these may have merged together, forming bigger black holes. Over time, through more merging, these black holes will have grown larger and larger and eventually sunk to the bottom of the potential well – the centre – of young galaxies.

Individual galaxies can also merge together if they are close enough, coalescing into one system. Eventually their central black holes will also merge, growing even more. As we know, a central black hole also feeds periodically on interstellar matter, maybe shining as a quasar. Over time, a supermassive black hole is grown, like a pearl within an oyster.

This remains speculation of course. Direct observation of Population III stars and primordial black holes is currently beyond us, although we are getting close to being able to detect the observational signatures of surviving Population III stars in very distant galaxies. Regular merging of the first black holes should also release a cacophony of gravitational waves, producing a chaotic 'background' of gravitational radiation. Now that we can actually detect gravitational waves directly, it might be possible to observe this background.

So the Population III stars would have been some of the light sources responsible for actually 'doing' the reionization at early times. But there's another way we can study this process: not through the light of the first stars and quasars, but through the electromagnetic radiation emitted by the neutral hydrogen gas itself. Like so many of the astrophysical processes we have encountered, despite the truly cosmological scale we are considering, we turn to quantum mechanics – the physics of the smallest scale – to describe how this light is emitted.

The classical picture of an atom is a dense central nucleus comprising protons and neutrons bound together by the strong nuclear force. The lightweight, negatively charged electrons, attracted to the positively charged nucleus, orbit like planets in some tiny Solar System. This isn't such a bad place to start, but quantum physics gives us a much better picture of the structure of the atom, specifically in the distribution of electrons around the nucleus.

The electrons are not haphazardly distributed around the nucleus but reside in discrete levels of increasing energy, which we can visualize as a set of shells around the nucleus. Electrons can make transitions between energy levels, either absorbing or emitting energy in the form of a photon, and that photon energy corresponds to the size of the gap. But the electrons aren't really like planets orbiting a star at different distances. As we have learnt, a quantum weirdness is that the electrons are not distinct entities at a definite point in space: until it is observed, a quantum particle only has a certain *probability* of being at a certain location. The distribution of that probability in space and time is described by the wave function. This means the electrons in their different

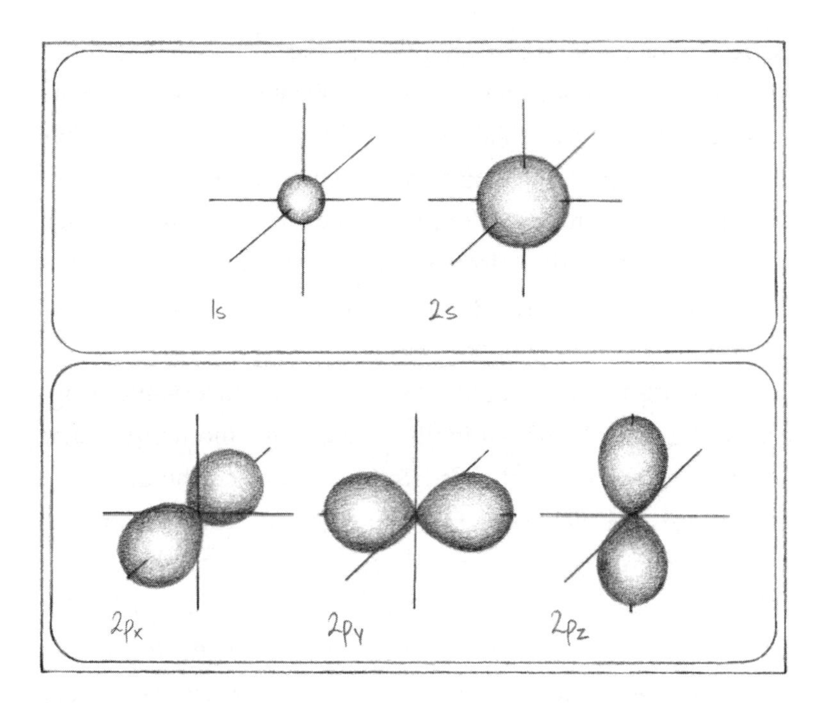

Electron orbitals

In quantum mechanics, electrons are not single point-like entities circling the nucleus. Rather, they are described in terms of the probability of existing at a particular point in space around the nucleus, as determined by solutions to the Schrödinger equation. Electrons can reside in different 'orbitals' depending on their quantum 'state', part of which is defined by the electron energy. No two electrons can be in the same state. The shape of the first five orbitals of hydrogen are shown here. The '1s' orbital is the lowest energy configuration and describes the 'ground state' of a neutral hydrogen atom.

energy levels around the nucleus can be thought of as fuzzy clouds rather than discrete particles. The shape and density of the clouds determines where the electrons are most likely to be.

Just as a pencil balanced on its tip will tend to fall over, in general electrons will try to eventually find the lowest energy state they can. So why don't all the electrons around atoms just emit photons and pile up in the lowest energy shell? The answer lies in a fundamental rule in quantum physics that we have already met: no two 'fermions' (an electron is a fermion) can occupy the same quantum state. This is the Pauli Exclusion Principle that also gives rise to the degeneracy pressure that supports white dwarfs and neutron stars. If one of the states in an atom is already occupied, then another electron cannot encroach on that state. So the electrons in the higher-energy shells cannot just arbitrarily transition to a lower energy.

A quantum state is defined by the set of observable properties of a particle, which include its energy, momentum and another property called 'spin'. Spin is a bit like the quantum mechanical equivalent of angular momentum in classical physics, and, like energy, it is also quantized – it can only take certain values. At a basic level, you can think of it as a kind of label for a particle. All electrons have a spin value of one half, but this can be oriented 'up' or 'down'. We call that a 'vector' quantity because it has a value and a direction.

Protons in the nucleus also have spin, and they too can be oriented up or down. In a hydrogen atom, where the nucleus is just one proton surrounded by one electron, there can be two spin-spin configurations: the proton and electron spins are aligned (up-up or down-down), or they are anti-aligned (up-down or down-up). Now, it turns out that in the configuration where the spins are aligned, the electron has ever so slightly more energy than the case when the spins are anti-aligned. The difference in energy corresponds to a tiny split in the ground energy level of the electron. We call it 'hyperfine splitting' because it is such a small energy difference.

A hydrogen atom that is in the slightly higher energy state can return to the lower energy state if the electron flips its spin. In doing so, it makes the hyperfine transition in energy, but just as in other atomic transitions, the electron must lose the extra energy by releasing a photon. In this case the energy gap is very small, so the outgoing photon has a correspondingly low energy, which means it has a long wavelength and low frequency. For neutral hydrogen this frequency is very close to 1,420 megahertz, or a wavelength of 21 centimetres, in the radio wave part of the electromagnetic spectrum.

The hyperfine transition should be known to other intelligent observers in the Universe, because they are likely to have also discovered quantum physics. This was the thinking when a diagrammatic representation of the spin-flip transition was chosen as one of the figures on the famous Pioneer plaques; engraved aluminium plates attached to the spacecraft Pioneer 10 and 11. The Pioneers were launched in 1972 and 1973, sent out to survey

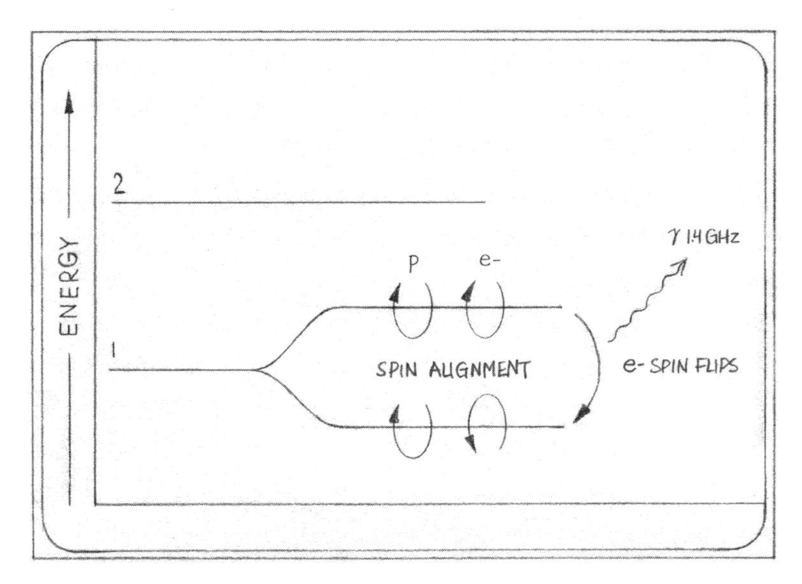

Hyperfine splitting

Hyperfine splitting of the ground state of the hydrogen atom. The energy difference between the spin aligned and anti-aligned configurations corresponds to a photon with a frequency of 1.4 gigahertz, or a wavelength of 21 centimetres.

the deep Solar System, eventually heading out into interstellar space. We have lost contact with the craft, but they are still out there, presumably hurtling through the lonely purgatory of the interstellar medium.

Each spacecraft carried diagrams representing humans and our place in the Universe, should either probe be intercepted by another civilization. In addition to the hyperfine transition, the plaques depicted a male and a female human body, our Solar System, the relative location of the Sun with respect to the centre of the galaxy and the distances and periods of fourteen pulsars (spinning neutron stars that regularly pulse radio waves) that lie close to the Sun. Our definition of time in units of seconds is rather arbitrary, so the pulsar periods were given in units of the inverse frequency of the hyperfine transition, which is a universal.

Although the hyperfine transition is a natural event, it is exceedingly rare to spontaneously occur for any given atom. We call it a 'forbidden' transition. In fact, if you watched a single atom of hydrogen in the state where the electron and proton spins were aligned, you would have to wait on average ten million years for the electron to spontaneously flip its spin and release a radio photon. But in astrophysics, we have the luxury of large numbers. There is so much neutral atomic hydrogen in the Universe that clouds of the stuff actually glow quite prominently with 1,420-megahertz radio waves; at any one time there are plenty of atoms undergoing the hyperfine spin-flip transition, either spontaneously or through certain triggering mechanisms like particle collisions.

With a radio telescope tuned to exactly 1,420 megahertz, we can map the neutral atomic hydrogen gas in and around our galaxy and beyond. In the Milky Way and in similar galaxies nearby, the stellar disc is laden with neutral hydrogen, and the tenuous gas extends far into the galaxy outskirts. When we look at spiral galaxies in radio waves, the neutral hydrogen appears like a cloak around the stellar pinwheel. This provides an excellent tracer of a galaxy's outer structure. It was this cloak-like feature of neutral atomic hydrogen that provided some of the first evidence that dark matter exists in and around galaxies. The clue was in the rate at which the gas is orbiting the galaxy.

In a spiral galaxy like the Milky Way, the stars in the disc are rotating around the central bulge in huge circular orbits. The Sun makes one complete revolution around the centre of the galaxy every two hundred and fifty million years or so, travelling at a circular speed of a little over 200 kilometres per second. The rate at which we orbit the hub is determined by our distance from the centre of the galaxy and the total amount of mass enclosed by our orbit, just like the rate at which the Earth orbits the Sun depends on the Earth–Sun distance and the mass of the Sun. In the case of the galaxy, the mass internal to our galactic orbit is not all in one point like it is in our Solar System, but spread out over many stars and gas clouds.

In other galaxies we can measure the rotation speed of stars and gas in the disc through spectroscopy: we can see the small changes in the observed frequency of bright spectral emission lines across the galaxy that correspond to Doppler shifts. For rotating disc galaxies oriented edge-on to us, we see the light from one side of the galaxy slightly blueshifted (coming towards us) relative to the other side, which is slightly redshifted (going away from us). By mapping the blue- and redshifts, we can measure the rotation speeds of other galaxies by translating the Doppler shifts in the frequency of light to the corresponding velocity shifts required to produce them. With sensitive, high-resolution observations we can do this even for galaxies seen far back into cosmic time. And since the rotation speed is determined by the amount of enclosed mass, we can use the pattern of rotation across a galaxy to measure the distribution of mass within it.

We can measure the distribution of *visible* mass in the galaxy by looking at the distribution of stars and gas that emits (or absorbs) electromagnetic radiation. For a spiral galaxy, most of the visible mass is concentrated in the bulge and disc, petering out as you go out into the outskirts of the galaxy. Based on the distribution of visible mass, the speed of rotation is expected to be slower towards the very centre of the galaxy, rising quickly as you go out in galactic radius and then slowing down again as you reach the edge of the stellar disc. We call the run of the speed of rotation with galaxy radius the 'rotation curve'.

In the late 1970s and early 1980s, Vera Rubin, an astronomer studying the rotation speeds of nearby spiral galaxies using one of the principal spectral emission lines of ionized hydrogen, noticed something odd. When the rotation speeds were plotted against the galactic radius where they were measured, the rotation curve stayed fairly flat, rather than declining as was expected.

The diffuse nature of neutral atomic hydrogen allows you to probe much further out into the disc than the stars and ionized gas, so you can measure the rotation curve to much larger scales than is possible with emission lines in the visible part of the electromagnetic spectrum. Observations of radio emission from neutral atomic hydrogen far out in the disc showed the same flat rotation profile. These flat rotation curves indicated that there is additional mass present in the galaxy not accounted for in any electromagnetic census. Moreover, the shape of the profile suggested that the additional mass must have a roughly spherical distribution – a 'halo' within which the flat stellar disc is embedded, like a fly in amber.

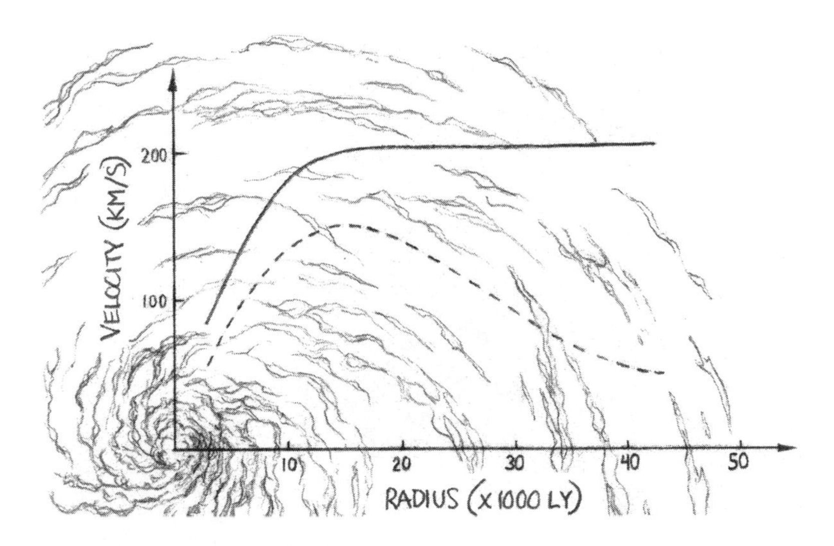

Galaxy rotation curve
Spiral galaxies like the Milky Way rotate like a spinning plate. The rotation speed changes with radius from the centre, and depends on the total amount of mass enclosed by that radius. We can chart this as a 'rotation curve'. A flattening of the rotation curve at large galactic radius was some of the first evidence for the presence of a dark matter component to galaxies.

This was some of the first observational evidence for dark matter, which we think makes up nearly a quarter of the total mass–energy density of the Universe. It is invisible but betrays its presence through gravity, affecting the motion of the normal, visible matter. Although we haven't yet directly detected dark matter (it is thought to be some type of particle), the picture of galaxies growing within dark matter halos has become the standard paradigm for galaxy formation models.

At the time of recombination, a few hundred thousand years after the Big Bang, the normal matter and dark matter were mixed together in a fairly uniform soup, rippled with density fluctuations. As time went on, gravitational amplification and collapse of these fluctuations formed the first halos of dark matter and

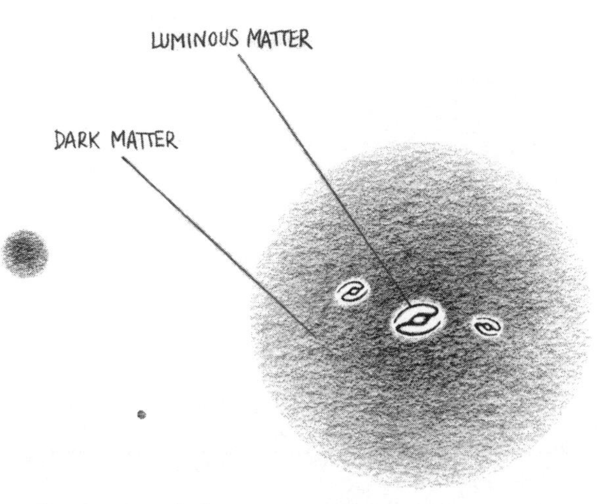

Dark matter halos
The luminous parts of galaxies – made of normal baryonic matter – sit within 'halos' of dark matter, which (as far as we can tell) only interacts with normal matter gravitationally.

teased the neutral hydrogen in the intergalactic medium into filaments and dense clumps. Cosmic structure began to emerge. Then, like Swiss cheese, the neutral intergalactic medium was punched with holes, where ionizing radiation shone out from the infant galaxies. So instead of thinking about the sources of ionizing radiation, what if we think about the light being emitted by the gas itself? After all, atoms in that sea of neutral hydrogen should be undergoing the hyperfine transition, releasing radio waves. Radio waves from the cosmic dawn.

Like all light emitted by distant sources, photons with a frequency of 1,420 megahertz emitted at the time of reionization undergo cosmological redshift, losing energy on their long journey across the Universe. If we want to detect them, we must tune our instruments to correspondingly lower radio frequencies. For neutral hydrogen gas undergoing the hyperfine transition around the time of reionization, the light arriving at Earth today has a frequency of about 140 megahertz, redshifted to around a factor of ten lower than the 'rest-frame' value of the hyperfine transition. This low-frequency radio signal is also extremely weak, due to cosmological dimming: distant sources of light are also very faint. Until very recently we lacked the technology to detect this radiation, because we did not have radio telescopes operating at sufficiently low frequencies sensitive enough to detect it. That's changing now.

A new generation of radio telescopes is coming online that aim to detect the radio waves emitted by neutral atomic hydrogen gas during reionization. The problem is that the radio signal from the Epoch of Reionization is not only tiny, requiring significant investments of hundreds or thousands of hours of observing time just to detect it, but it is also swamped by other radio signals in the Universe. All of these contaminating signals are emitted by structures that formed well after the Epoch of Reionization, and so from our point of view appear as 'foregrounds' of light, blanketing the primordial radio signal we're interested in.

Galaxies themselves emit radio waves. As well as the glare of visible, ultraviolet and X-ray light from quasars, actively growing black holes can drive narrow jets of particles at near the speed of light via extreme magnetic fields. These jets often penetrate right through the galaxy and far into intergalactic space, pluming like cigarette smoke. Electrons in the jets are accelerated by magnetic fields, and this interaction releases radio waves called 'synchrotron' radiation. Powerful 'radio galaxies' are relatively rare (they tend to be the most massive galaxies) but they can be so bright in images taken with a sensitive radio telescope that they will completely obliterate any faint background signal within their vicinity, effectively photobombing deep radio maps.

Through a similar physical process, all galaxies that are forming stars tend to emit a broad continuum of radio light, but at a much lower level than the powerful radio galaxies with active black holes. The sporadic explosions of supernovae, occurring maybe a few times per century for a typical galaxy, also accelerate electrons through the interstellar medium. Galaxies are laced with magnetic fields, and when the electrons encounter those fields, their paths are deflected. It is this interaction between charged particles and magnetic fields that releases photons in the radio wave part of the spectrum. Although relatively faint, star-forming galaxies are numerous throughout the Universe and are therefore another contaminant to consider when attempting to detect the faint Epoch of Reionization signal. To cap it all off, we are trying to observe the deep Universe *through* our own galaxy. The Milky Way is full of different sources of radio waves, and because we are in such close proximity to them, they are extremely bright compared to the distant signal we are after.

One of the brightest radio sources outside the Solar System is called the Crab Nebula. The Crab is what remains of a star that exploded in 1054. We know the year because the explosion, or at least its immediate aftermath, was actually observed by Arab and Chinese astronomers: the supernova appeared as a new star in the sky, and was dutifully recorded. Today the remnant of that dead star is a pulsar – a rapidly spinning neutron star that emits narrow beams of radio waves. When the beam intercepts Earth we see a short pulse of electromagnetic radiation, like a lighthouse. The pulsar also drives a highly energized wind of gas into the wider nebula, the shattered remains of the star that once was. As a result, the pulsar and the surrounding gas blaze with radio waves. Living inside a galaxy really is a nuisance if you are interested in studying the wider Universe: you can't see the wood for the trees.

The foreground galaxies emit a radio signal about a thousand times brighter than the reionization signal we're interested in, and the Milky Way itself emits a signal about a thousand times brighter still. So this observational challenge is like trying to isolate the ripples from the splash of a pebble that has been thrown into a stormy sea.

But there is good news. Like finding a break of blue sky on an overcast day, we can look in directions that have the lowest levels of contaminating foreground emission. Invariably, these are regions of the sky that avoid the dense plane of our own galaxy, which is so thick with gas and stars that it is impossible to see the Universe beyond. Instead, we can find the gaps through the galaxy, away from the disc, where there is a little less intervening material. These windows give us a clearer view of what lies beyond. This strategy is true of all extragalactic astronomy. Still, in order to detect and map the reionization signal, you do need to observe a reasonably large swathe of sky, and inevitably foreground contamination from our galaxy and other galaxies remains. It must be removed to reveal the tiny background signal.

Luckily, the nature of the hyperfine transition helps us out here. Most of the contaminating foreground signal is emitted over a broad range of radio frequencies, with a smoothly varying spectrum. The signal from the neutral hydrogen is, in contrast, a distinct spike of emission; the hyperfine transition occurs at a very specific frequency, even if that frequency is redshifted by the time it reaches our telescopes. It is this difference in the spectral properties of the radio emission from different sources that allows us to filter out much of the contaminating signal. Of course, I have glossed over the nasty technical details, not doing justice to the clever design of the telescopes and receiver technology, or the observations themselves, or to the data processing and cleaning of the signal, or to the complex scientific analysis and interpretation itself. There are myriad challenges to overcome.

But these challenges *will* be overcome. Telescopes such as the Low Frequency Array, a network of low-frequency radio receivers spread across Europe, and the new Canadian Hydrogen Intensity Mapping Experiment in Penticton, British Columbia, among others, are now attempting to observe the radio light from neutral hydrogen during the Epoch of Reionization.

The basic goal of these experiments is not just to detect the radio signal, but also to measure fluctuations in the signal across the sky. The exact distribution of fluctuations in the intensity of the light can be compared to theoretical predictions for different models of how reionization proceeded, given our current understanding of cosmology and galaxy formation. This comparison between data and theory will advance our knowledge of the early Universe.

If the neutral atomic hydrogen gas was totally smooth and completely filled the Universe, then we should see no variation in the power of the radio signal on different scales. But the reality is that the gas is clumpy: it traces cosmic structures as they are forming under gravity, and therefore the density of the gas changes from place to place. The number of photons being emitted through the hyperfine transition is correspondingly stronger or weaker depending on how much gas is there. More importantly, as reionization proceeds, we can think of the volumes around new galaxies and quasars becoming increasingly ionized.

With no neutral hydrogen there can be no hyperfine transition, and so we expect the radio signal to weaken and go dead in the immediate vicinity of the sources of reionization. The pattern of how this happens will be imprinted on the radio light, and therefore – if it can be detected – we have a unique way of observing when reionization occurred, how long it took and how it unfolded across the Universe as the first galaxies lit up.

The study of galaxy formation is like reading a novel backwards. We started with the final pages, detecting the light from nearby objects that has taken a relatively short time – perhaps a few million years or less – to reach us. This is the stuff that is easy to observe: our own galaxy of course, and our galactic neighbours, like the Andromeda Galaxy. They are the product of nearly fourteen billion years of cosmic evolution. The astounding thing about extragalactic astronomy is that we can actually watch this evolution happening. Simply by collecting the light that has taken perhaps billions of years to reach us, we can understand what galaxies were like in the past.

As we peer further back in space and time, astronomical sources become extremely faint, and so we rely on fewer and fewer photons to understand the story. It's as if we find more missing letters the further we read. Worse, the photons with which we read the story are almost lost in a Universe that is dazzling with electromagnetic radiation from countless sources. Yet despite the technical challenges involved in collecting the light, and the need to devise increasingly canny techniques to decipher its message, we have now observed nearly all stages of galaxy formation in the Universe. Except the very start of the story. Detecting the radio light emitted from the cosmic dawn is like reaching the first chapter of the novel. It will tell us how galaxies began.

EPILOGUE

Astronomy is at once a frustrating and exhilarating science. We cannot travel any meaningful distance from Earth, and while our descendants may one day visit the nearby stars, it is doubtful our species will ever explore and inhabit the galaxy as a whole in the same way humans have the Earth. And it is certain we will never travel to other galaxies.

Even if we could do these things, humans are like mayflies to the cosmos. If the age of the Universe was one year, the entire time span of all human civilization amounts to no more than the last thirty seconds of the clock. A single life is the blink of an eye.

So we can never fully appreciate the growth of cosmic structures, or the slow heave of interstellar gas, or the gravitational waltz of two galaxies as they collide, or the birth and death of stars. In our one-year-old Universe, most astrophysical processes would take place on timescales lasting hours and days, weeks and months. We see but a snapshot, a single frame, when we look out into space.

And yet, we *can* understand these things, simply by looking out and collecting the light that has journeyed across the cosmos. We read from it the story of the Universe and its contents. Without having been there, we know what the Universe was like shortly after it came into existence, and how the seeds for the formation of galaxies were imprinted. We can detect the expansion of the Universe and its apparent acceleration. We can map gravitational distortions in spacetime itself, simply by detecting the deflection in the paths of photons as they pass through it. We can imagine the ten-million-year escape act of a photon from the core of a distant star, and observe how galaxies have changed over cosmic history. We can measure the internal motions of those galaxies and determine their composition. We can delve deep into their hearts and picture ourselves at the edge of a supermassive black hole that is gobbling up the interstellar medium, releasing unfathomable amounts of energy. We can see how these objects shine across the Universe. We are now seeing light from the cosmic dawn.

Our curiosity – what drives us on – is the same as that of our ancestors: a simple craving to know and try to understand. Science slakes that thirst. Sometimes the endeavour can seem esoteric and indulgent: should we be spending millions, or billions, of taxpayers' money to build instruments that allow us to see a little bit further into the Universe, to know a little more about how it works? Where is the tangible return on that investment? How does it make your everyday life better?

The truth is, in the very short term, it probably doesn't, except in the same way that a new song or piece of art or blockbuster movie makes everyday life better. A new discovery might make a splash in the news, accompanied by beautiful graphics and enthusiastic missives from the scientists involved. It may catch your eye, giving you a few moments' wonder before you move on with your day. Fundamentally, and of true value, is the fact that humans have learned something new about the Universe, and we mature as a civilization as a result. And the real societal dividends come later on.

There are the children who are inspired to become scientists and engineers in many different fields. They will invent the technology of the future and cure tomorrow's diseases. Then there are the creative solutions to complex problems that find application in many other fields, like the signal processing algorithms that allow wireless networking to work properly in your home, which were originally developed by radio astronomers seeking to understand black holes. Or the link between adaptive optics technology that delivers both sharper astronomical images and also improved retinal scanning to detect eye disease. These things slowly trickle into society without most people realizing it. And while the practical benefits of fundamental scientific research can be unpredictable, they are often profound.

That we choose to try to figure out the workings of the Universe is the reason we live in the Space Age, not the Stone Age. Provided we continue this quest of understanding, there is the promise that the technological leaps to be expected in the coming centuries will, in the eyes of our descendants, make our twenty-first-century civilization appear as medieval society does to us.

Despite how far we have seen, still we strain our eyes, eager to see more. I think we always will.

FURTHER READING

Al-Khalili, Jim, *Quantum: A Guide for the Perplexed* (London, 2003)
Arcand, Kimberly, and Megan Watzke, *Magnitude: The Scale of the Universe* (New York, 2017)
Coles, Peter, *Cosmology: A Very Short Introduction* (Oxford, 2001)
Cox, Brian, and Jeff Forshaw, *Universal: A Journey through the Cosmos* (London, 2017)
Einstein, Albert, *Relativity: The Special and the General Theory* (Princeton, NJ, 2015)
Feynman, Richard P., *The Character of Physical Law* (London, 1992)
—, *QED: The Strange Theory of Light and Matter* (London, 1990)
—, *Six Not-so-easy Pieces: Einstein's Relativity, Symmetry, and Space-time* (New York, 2011)
Galfard, Christophe, *The Universe in Your Hand: A Journey through Space, Time and Beyond* (London, 2016)
Geach, James, *Galaxy: Mapping the Cosmos* (London, 2014)
Green, Lucie, *15 Million Degrees: A Journey to the Centre of the Sun* (London, 2017)
Hawking, Stephen, *A Brief History of Time* (London, 1998)
Misner, Charles W., Kip Thorne and John A. Wheeler, *Gravitation* (Princeton, NJ, 2017)
Mo, Houjun, Frank van den Bosch and Simon White, *Galaxy Formation and Evolution* (Cambridge, 2010)
Sagan, Carl, *Cosmos* (New York, 2013)
Scharf, Caleb, *Gravity's Engines: The Other Side of Black Holes* (London, 2012)
Smoot, George, and Keay Davidson, *Wrinkles in Time: Imprint of Creation* (New York, 1993)
Susskind, Leonard, *Quantum Mechanics: The Theoretical Minimum* (New York, 2014)
Thorne, Kip, *Black Holes and Time Warps: Einstein's Outrageous Legacy* (New York, 1994)

Tyson, Neil deGrasse, *Astrophysics for People in a Hurry* (New York, 2017)
Weinberg, Steven, *The First Three Minutes: A Modern View of the Origin of the Universe* (New York, 1993)

ACKNOWLEDGEMENTS

It is a pleasure to thank my brother, Tim Geach, and my colleagues Elias Brinks, Martin Hardcastle, John Peacock and Douglas Scott for taking the time to read the manuscript and providing sage advice and suggestions for improving the text. It was a delight to work with Brett Harding – his beautiful illustrations bring to life many of the key astrophysical concepts I attempt to describe in the text. Finally, I would like to thank my wife, Kristen. Without you, I'd be lost in a lonely cosmos.

INDEX

Page numbers in *italics* refer to illustrations